Blind Spot

BLIND SPOT

WHY WE FAIL TO SEE THE SOLUTION
RIGHT IN FRONT OF US

How Finding a Solution to One of the World's Greatest Mysteries
with the Verifier Method Changes the Way We Approach Success

GORDON RUGG

with

JOSEPH D'AGNESE

HarperOne
An Imprint of HarperCollins*Publishers*

HarperOne

All Voynich Manuscript images courtesy General Collection, Beinecke Rare Book and Manuscript Library, Yale University.

HarperCollins books may be purchased for educational, business, or sales promotional use. For information, please e-mail the Special Markets Department at SPsales@harpercollins.com.

HarperCollins website: http://www.harpercollins.com

HarperCollins®, ®, and HarperOne™ are trademarks of HarperCollins Publishers.

FIRST EDITION

Library of Congress Cataloging-in-Publication Data
Rugg, Gordon.
    Blind spot : why we fail to see the solution right in front of us :
how finding a solution to one of the world's great mysteries with the
verifier method changes the way we approach success / Gordon Rugg
& Joseph D'Agnese. — First edition.
        pages   cm
    ISBN 978-0-06-209790-3
    1. Problem solving.   2. Expertise.   3. Knowledge, Theory of.
I. D'Agnese, Joseph.   II. Title.
    BF449.R84   2012
    153.4'3—dc23

                                                            2012046662

13 14 15 16 17   RRD(H)   10 9 8 7 6 5 4 3 2 1

Contents

Introduction   1

1. The Expert Mind   11

2. The Knowledge Within   29

3. The Imperfect Expert   49

4. From Words to Images   69

5. Ambush in Your Mind   101

6. The Voynich Manuscript   123

7. Dissecting Verifier   153

8. The Mathematics of Desire   179

9. Fuzzing into Mush   209

10. Joining the Strands   243

Acknowledgments   281

Notes   283

Index   289

Introduction

LOCKED AWAY IN A LIBRARY AT YALE UNIVERSITY IS a mysterious medieval book full of bizarre illustrations, its words handwritten in a unique script that has never been deciphered. The man who discovered that book in 1912, in a seminary in Italy, was an antiquarian bookseller named Wilfrid Voynich, a former anti-tsarist revolutionary who fled Russia after some of his friends were killed by the secret police.

Voynich knew a lot about secrets, and he immediately understood the significance of the mysterious book. It was probably written either in code or in an unusual language; either way, it would be extremely valuable. He bought it immediately, and anxious to read its contents, he did what anyone does who has a tough problem to solve: he asked an expert to help. When the best minds in cryptography failed him, he turned to others and still others. They got nowhere. Voynich died, but the book's reputation lived on, tantalizing fresh generations of expert code breakers with its secrets.

Ninety years later, the book was still undeciphered. By then, it held an almost iconic status. It had become a rite of passage for budding young cryptologists, who bashed their brains against the book, failed, and moved on to more feasible projects. Even the men who cracked the codes of the Nazis and the Japanese during World War II failed to unlock the secrets of what is now known as the Voynich Manuscript.

Undeciphered codes aren't just good for romantic, mysterious stories. They have implications for everyday life. The Internet depends on codes; without trustworthy codes, online commerce would grind to a halt, along with large swaths of national security. Modern code breakers are very, very good. There's a problem, though. Even the best modern codes are reaching the end of their shelf life. It's possible to crack them if you have a lot of computing power and don't mind waiting months for a solution. For the moment, that sort of crack isn't worth it. In a few years, however, technology will likely make that crack in seconds rather than months. The Voynich Manuscript could perhaps help cryptographers to design new, unbreakable codes—*if* anyone could figure out what it meant.

I began working on the Voynich Manuscript in 2002. Within a few weeks, using just a pen, ink, and replica parchment, I discovered something completely unexpected. I found compelling evidence that the writing in the book was neither a language nor a code but a grandiose hoax—a glittering artifact containing nothing but gibberish, probably created by a con man for the sole purpose of bilking a monarch out of money. The evidence was so strong that I was able to publish it in a peer-reviewed academic journal—the first proposed solution for the Voynich Manuscript to get through peer review in almost half a century.

I'm not a code breaker by trade. I'm a psychologist and a computer scientist, though I didn't use a computer to find what was lurking in the Voynich Manuscript.

The story of my discovery was picked up by media all over the world. That wasn't surprising: the story made good copy, with its references to some of the most colorful characters in sixteenth-century Europe, including the Holy Roman emperor and Queen Elizabeth I, as well as one of the greatest con artists of all time and a man known as the most degenerate cleric of the age. The story even featured a treasure map, but most journalists didn't bother to include it, because juicier parts of the story jostled for space.

My proposed solution—which is still hotly debated by dedicated Voynichologists around the world—looked surprisingly simple. How could it be that experts in mathematics, code breaking, languages, and other important fields had missed something that an outsider was able to spot? Simple. Those experts made a mistake.

Though I'm not an expert in code breaking, I *am* an expert in human error, specifically the errors of experts. In my line of work, we study how humans acquire expertise, and we know that when experts screw up, they screw up in predictable ways. If you know how to look for these mistakes, you can easily spot where experts went wrong. For ninety years, the experts looking at the Voynich Manuscript had operated with a blind spot. They were focusing on the things they thought were important but ignoring possibilities that were right in front of them the whole time.

The Voynich case, nice as it was, was not my endgame. It's a lovely problem, but my colleagues and I were after a larger quarry. Specifically, I wanted to test a theory I had been working on for years. The Voynich Manuscript was only the first test of a toolkit of ideas, theories, and strategies my colleagues and I call the Verifier Method. We think Verifier is a powerful new contribution to science, suitable for checking the work of experts in any field. It's the scientific equivalent of taking on the hardest cold cases in detective work. The method is designed specifically for problems where researchers probably already have the relevant information but can't make sense of it.

In my line of work, we are interested in how humans acquire, store, and use knowledge. Researchers like me have become very good at figuring out how to represent, visualize, or model the knowledge contained in human brains. This started for a practical reason: we needed to know what humans knew so we could program computers to help them do their work. Paradoxically, perhaps, the more information we got from humans, the more we began to see how special human brains were.

Our first Verifier studies showed us that one of those special human gifts—which we'll discuss at length—is a key tool for solving so-called intractable problems. This should not have been a surprise. Computers are solid, logical plodders, good at performing rigorous, complex, methodical tasks that would stress most humans. We humans, on the other hand, are good at swift, nimble, cut-to-the-chase thinking that is strongly governed by visual cues. In the course of this book I'll argue that if you want to solve complex problems, you need to know which tools—and which brains—to use.

The Verifier Method is based on a large collection of tools. Those tools include software, problem-solving strategies, and more, all built on a bedrock of formal logic. We've used this method to look at problems more important than a mysterious book. Medical problems. Criminal cases. Problems in our understanding of the cosmos and the world we live in. The findings are tantalizing; it looks like we're onto something.

This book is about the birth of that method, told as a sort of intellectual detective story. As we delve into that story, we're going to throw some challenging concepts at you, which you'll need to understand before moving on to the next pieces of the puzzle. For now, it's important that Verifier is only one component in a larger framework for understanding and working with human knowledge, best expressed by Figure 1. This image reflects the structure of this book.

The first third of our story is about *unpacking* knowledge from the minds of human experts. Scientists like myself perform such unpacking to design software that helps experts do their work. Our tour through this landscape will introduce you to some colorful characters. Card sharks. Code breakers. Chess

The Knowledge Cycle

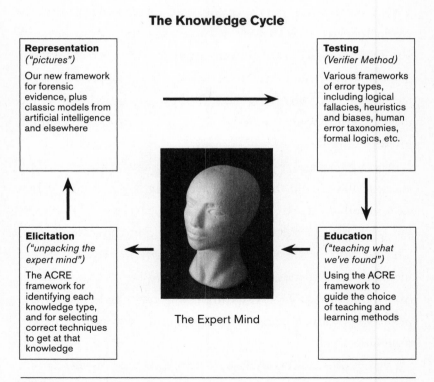

Representation
("pictures")

Our new framework for forensic evidence, plus classic models from artificial intelligence and elsewhere

Testing
(Verifier Method)

Various frameworks of error types, including logical fallacies, heuristics and biases, human error taxonomies, formal logics, etc.

Elicitation
("unpacking the expert mind")

The ACRE framework for identifying each knowledge type, and for selecting correct techniques to get at that knowledge

The Expert Mind

Education
("teaching what we've found")

Using the ACRE framework to guide the choice of teaching and learning methods

FIGURE 1

masters. Swordsmen. Brewmasters. We'll get to meet that gullible emperor and the sociopathic con man, both of whom lived in the days of Shakespeare. We'll even meet the wizard-like astrologer who advised Queen Elizabeth I. We'll look in on all these people, each an expert in a rarefied field—even the con man!—and see what each has to teach us about human expertise. Specifically, we'll learn that expertise has limitations. Just because you know something doesn't mean you can put it into words, teach others, or tell others how you're thinking about a problem.

If a scientist is clever, she will use several techniques—verbal techniques, visual techniques, or tools—to elicit the truth from experts. Because human beings are strongly visual, the scientists we'll meet on our journey use one of the oldest tricks in the world to extract and transmit knowledge: they draw pictures and rely upon our natural human gift for sizing up things at a glance. The second part of the book talks about *visualizing* or *representing* that knowledge, and I'll show you how I used visual tools to get at what I consider to be the truth of the Voynich mystery.

The clues to any mystery are bound up in the ways we think about problems and the pictures we create in our heads. When experts fail to solve a problem, it's often because of a blind spot: the truth is right there in front of them, but they don't see it. The pictures in their heads conflict with the reality they're seeing or *think* they're seeing. How do we help them break out of that blind spot? Well, you can use a system like Verifier to *test* or *check* their work, and then *teach* them how to problem-solve more effectively.

We got good results studying difficult problems with Verifier. We realized that problem solving works best when you let humans do what they do best—visually scan large chunks of data and results, then use their amassed knowledge of a problem to arrive at a solution. The need for tools to enhance visualization led us to develop software to aid this process. One program we developed, the Search Visualizer, lets you whip through results of a web search even if you don't speak the language of the material you're "reading."

I'll show you what I mean about all this—visual tools, pictures in your head, and the conflicts at the heart of any blind spot—in the last third of the book. But I might quickly hint at the big picture by sharing the story of stomach ulcers. It's a great story about a blind spot that killed people.

Stomach ulcers are a particularly unpleasant way to die. For centuries, well into the twentieth century, doctors couldn't do much beyond prescribing antacids and recommending a lower-stress lifestyle with a carefully managed diet. Then a couple of Australian doctors challenged the orthodox view of ulcers by claiming that stomach ulcers were actually caused not by the body's reaction to stress, but by a bacterium living in the human stomach. That idea got a pretty negative reception, for sensible reasons. The human stomach is an extremely hostile environment, with acidity levels that would kill any known bacterium.

Within a few years, though, the two Australians demonstrated that a bacterium could survive in the human stomach, that it did cause a significant proportion of stomach ulcers, and that those ulcers could be successfully treated with standard antibiotics. The bacterium was *Helicobacter pylori*. The Australians were Barry Marshall and Robin Warren. Their discovery saved thousands of lives, and they won a Nobel Prize for it. It's a classic case of a revolutionary new insight that changes the world for the better.

There wasn't a significant mistake in what Marshall and Warren did—quite the contrary. The mistake was in what large numbers of other medical researchers *didn't* do. Marshall and Warren weren't the first to see *Helicobacter pylori*. It had been spotted by at least three separate teams of researchers a century earlier, and there had been at least one suggestion that it was implicated in stomach ulcers and gastritis (irritation of the stomach). Yet no one picked up that idea and ran with it a century earlier.

That's the sort of problem that catches my attention. You can't point at an individual decision and identify it as the place where things started to go wrong. It's harder than that. It's not that we don't yet have the technology or the data to answer the key question. It's more complicated than that. Sometimes it involves finding the right way to look at a problem. Sometimes it involves explanations that are nearly, but not quite, right.

A "nearly right" explanation for something can pass unnoticed for a long time, causing damage in science and in the real world without getting spotted. A "wrong" explanation is easier to spot, and tends to get fixed quicker, if a fix is possible. At the very least, a wrong theory gets a warning sign posted in front of it, if it can't be fixed.

I'm interested in the problems where we already have all the pieces of the puzzle, but nobody has worked out how to fit those pieces together and give us the final picture. Visible mistakes are easy to spot. A bridge collapses, or an aircraft crashes, or a company goes bankrupt, because someone made a series of misjudgments. There's been a lot of scientific work on visible mistakes. Other mistakes are harder to see. This is, at heart, a book about invisible mistakes made by conscientious professionals hampered by their own blind spots.

Why do such mistakes happen? Sometimes it's because nobody sees that all those pieces are part of the same puzzle. Sometimes it's because everyone has made the same wrong assumption about how some pieces fit together. Sometimes it's because the professionals involved don't have the training to push through the logjam. All science-based researchers these days are trained to use two tools—statistics and research methods. But they're rarely taught about human error, even though a key goal of research design is to minimize the risk of flawed findings caused by human error.

Attempts to prevent human error are nothing new. Logicians have been pointing out flaws in human reasoning for over two and a half thousand years, without much visible effect on everyday life; human error is still all around us. This time, though, the game might be changing. Previous attempts to fix this problem tackled it by telling people what they should be doing—explaining the correct answer in logical arguments, one step at a time. That makes perfect logical sense, but it plays to people's worst weaknesses. We realized during our Verifier work that there's a very different way to tackle the same problem, by playing to human strengths that can also paradoxically be our weaknesses. It's exactly like the transition from old-style computer interfaces, where you typed in cryptic verbal commands, to present-day interfaces, where you click on icons and images. We explain later in this book just how this new approach works, and why it just might change the game forever.

Or it might not. But one thing is clear. In an age that's drowning in an ocean of information, we as a society need to reengineer the way we solve problems. Otherwise, there will be more tragically missed opportunities like the decades of failing to spot an effective treatment for stomach ulcers. Maybe

Verifier will be at the forefront of this new wave; maybe it will turn out to be a sideshow, with the main event coming from yet another new direction. However it all works out, there's now a realistic chance of solving long-unconquered challenges far more important and more mysterious than the strangest book in any library.

The Expert Mind

N THE SIXTEENTH CENTURY, EXCAVATORS DISCOVERED A glass vase in some ruins outside Rome. Roman glassworkers were good by anyone's standards, but this piece of glassware is exceptional. It's made of two layers of glass, a dark blue inner layer with a white outer layer. The surface is carved into rich, intricate cameos depicting humans and gods. It's known now as the Portland Vase, after the British aristocrat who bought it. It was made around a couple of thousand years ago. The glass-cutting style used to make the vase is so complicated, so intricate, that archaeologists think it probably took the original craftsman two years to make.

One way of measuring the skill of the vase's maker is to look at how long it was before anyone was able to produce another piece of glassware like it: it took almost two thousand years. Producing a replica became a challenge for the best glassmakers of the Industrial Revolution. The vase became such an iconic challenge that Josiah Wedgwood developed a whole line of Wedgwood pottery inspired by it. The Great Exhibition of 1851 was a high-profile showcase for the crowning achievements of the era—the most sophisticated technology in the world. It didn't contain a replica of the Portland Vase, because nobody had been able to produce one. The first passable-quality replica was made in the 1870s.

It was that difficult. That's one of the first lessons about experts.

LESSON 1
Experts can do things that the rest of us aren't able to do.

We're surrounded by examples of the complexity of expert knowledge in everyday life, to the point where we don't even notice them most of the time. Most offices have computer support staff; they are experts at doing things most of us don't begin to understand. On the way to the office, many of us travel by train. We wouldn't have a clue how to drive a train, maintain the tracks, or schedule service. If you drive to work, you're using a vehicle that requires expertise to service and repair; it's not advisable to try diagnosing and replacing a faulty fuel injection system using general knowledge and common sense.

Experts are good, often impressively good, and they can do things that nonexperts can't. However, that doesn't mean they're immune to making mistakes. If we understood how experts think, we could probably help them avoid those mistakes. But it turns out that's easier said than done. In this chapter, I'll try to explain why it's so hard to know what experts know.

The *Real* Lightsabers and the Unreal View of War

One area where there's been a long-standing tension between true expertise and mistaken expert opinion is the world of war.

Ever wondered what inspired the lightsabers in the *Star Wars* movies? George Lucas and his team were probably inspired by the name of the nineteenth-century light—as in, not so weighty—saber. This isn't a steampunk creation bristling with brass work, dials, and levers; it's simply the alternative to the heavy saber.

Learning how to fight with a full-weight saber was dangerous, so fencing instructors, treading an uneasy line between realism and injury, began using lighter sabers. They deliberately minimized injury at the expense of realism, but the rank-and-file cavalry continued to use full-weight heavy sabers in battle. That choice made it more difficult for the instructors to teach cavalrymen how to fight an opponent who was using more brute force than skill. And the soldiers, thinking their instructors were out of touch after years of working with light sabers, tended to be skeptical about how much real expertise the fencing masters had. The soldiers' experience on the battlefield, documented in numerous diaries of ordinary troopers, suggested that they were often right to be skeptical. So the counterpoint to the last conclusion about experts is cautionary:

LESSON 2
Experts' skills don't always correspond to reality.

For centuries, it was taken for granted by most people that expertise had some features that distinguished it. It was generally assumed that experts were better at pure logic than lesser mortals, and that they used this logic combined with their higher intelligence to solve problems. There was also a strong element of snobbishness, with an implicit assumption that "real" expertise was the province of white-collar professionals, with manual skills being excluded from the club of expertise. The chess game was often viewed as an archetypal demonstration of expertise in action: a good chess player can easily beat a weak chess player, so there's clearly some real expertise involved, and chess is an abstract, cerebral skill, unlike sordidly manual skills, such as glass-making or hacking someone to death on a battlefield.

When the French chess master François-André Danican Philidor played three simultaneous blindfold games of chess in 1783, this was hailed as one of the highest achievements of the human intellect. Expertise distinguished humans from lesser creation, as well as distinguishing upper-class intellectuals from the lower orders. It was a comforting view, which wasn't seriously challenged until the 1960s, when everything changed.

Crumbling Walls

One of the first challenges to this cozy belief came from research into chess masters. Beginning in the 1940s, a Dutch psychologist named Adriaan de Groot and his colleagues began investigating how chess masters actually operated, as opposed to how everyone assumed they operated. This led to an important finding.

LESSON 3
Experts are not significantly more intelligent than comparable nonexperts.

De Groot and his colleagues found, to their surprise, that chess masters weren't significantly more intelligent than ordinary chess players. Nor did they have a significantly better memory for the positions of chess pieces placed randomly on a chessboard. Their expertise turned out to be coming from a completely different source: *memory about chess games.*

What set chess masters apart was that they could remember enormous numbers of gambits, strategies, tactics, placements of combinations of pieces, previous examples, and on and on, from games they had played, games they had watched, or famous games in history that they had studied. Although they weren't good at remembering random arrangements of pieces on a chessboard, they were very good at remembering nonrandom arrangements, such as a particular configuration of a king and several supporting pieces. The number of such memories was staggering: a chess master typically knew tens of thousands of pieces of information about chess. Later, when other psychologists began studying the way experts thought, a remarkably similar picture emerged from research into other areas of expertise: what defined an expert always turned out to be the possession of tens of thousands of pieces of information, which typically took about seven to ten years to acquire.

LESSON 4
Experts retain huge numbers of facts about their areas of expertise.

These findings were brutally consistent, even for venerated prodigies, such as Mozart, who began playing music at age three and composing by age five. But if you looked at how long it took between his first compositions and the first of his compositions *that stand up to comparison with expert composers,* the period is about seven to ten years. This discovery gave researchers pause: maybe expertise was simply a matter of time, experience, and training. Maybe if you put in enough time working at something, you would eventually become an expert.

Another blow to the foundations of the old views again came from chess. The first modern computers emerged during the 1940s. Within a couple of decades, there were computer programs that could beat an average-level human chess player. Soon after, computer programs could beat masters, and then grand masters.

Everyone had expected chess to be one of the last bastions of human supremacy over dumb beasts and machines. It was more surprising that computers had a lot more difficulty with the apparently simple game of Go, the ancient Asian board game, than with chess. That's because skill at Go relies more on nonverbal spatial knowledge—where the tiny black-and-white pieces are located on the board—which is tough to program. That's a theme we'll encounter repeatedly throughout this story.

LESSON 5
Just because a human finds something difficult doesn't mean that it necessarily *is* difficult.

The lessons from this research into the minds of experts weren't lost on applied researchers and industry. In fact, this research was leading to the next inevitable step: taking what we had elicited from human experts and using it to help us program computers to work better with humans or to perform tasks too tedious or complex for humans. In this chapter, I'd like to show you what scientists like me did with what they learned. In some cases, what we knew about experts helped us enormously. In others, it left us wondering if we didn't need to know still more about how experts think.

In the 1970s psychologists such as Daniel Kahneman, Paul Slovic, and Amos Tversky conducted a campaign parallel to the expertise research, looking at the types of mistakes humans make. They found that for some problems, simple mathematical linear equations could outperform human experts. By that, we mean that a small piece of software could predict better than an experienced bank manager whether or not a particular customer would default on a loan, predicting it more accurately, more reliably, and much more cheaply. That finding ended up with a lot of middle-ranking bank managers losing their jobs and being replaced by software. If you apply for a loan today, the decision on your application will almost certainly be made by a piece of software. Similarly, if there's a suspicious pattern of activity with your credit card, it will probably be spotted by a piece of software, which will trigger a check by a human being to make sure that your card or its number hasn't been stolen.

That finding had implications for other areas, such as medicine. Researchers wanted to create software systems that diagnosed illnesses the way doctors did. But the researchers wondered if the systems they built could go beyond mere imitation and actually improve the rate of correct diagnoses for life-threatening illnesses. The early signs looked promising. In the early 1970s, scientists at Stanford University developed a program called MYCIN, which could diagnose some categories of infectious diseases more accurately and more reliably than expert human diagnosticians. A doctor had to answer a series of simple questions about a particular case in order for the system to produce a diagnosis. This was so successful that by the 1980s, expert systems were already outperforming human experts in a range of areas, and it looked as if a new era of software-supported medicine was about to dawn. But then reality got in the way, as a slew of problems emerged.

In the fall of 1986, I moved to Nottingham to begin working on a problem that was causing a lot of difficulty for expert systems. It's known as the "knowledge acquisition bottleneck," and I was to encounter it repeatedly in the years ahead. It's about extracting knowledge from human beings to put into the expert system. To write a robust piece of software, you need to base it on solid knowledge from the real world. Getting at that solid knowledge was turning out to be a harder task than anyone had anticipated.

In the old days, software was built using what is known as the waterfall model. The software developer would interview the client, and then draw up a document that specified in detail what the system would do. Once that agreement was signed, the software developer went away and built the software. The clients weren't involved in the build; when they put their signatures on the contract, they were committed to that plan, as irrevocably as a log going over a waterfall.

Developers got information out of the experts via the traditional interview, and most software developers believed that this worked just fine. If the client forgot to mention an important requirement during the interview, that was the client's problem, and the developer would pick up a further fee for fixing the problem. But when the software keeps having problems after delivery, or each new version has new problems, or the software simply doesn't do what the client wants, the limitations of this approach become obvious.

The first expert systems were built in one of two ways. Some

were built by people who had come into the field from traditional software development; they used interviews because interviews were the only approach they had ever known. Others
were built by what are known as domain experts: people who
were already experts in the relevant area and who had then
learned how to build expert systems. These people wrote their
own knowledge into the software, without needing to interview
anyone. Whichever route was used, the early expert systems
could perform better than human experts for tightly defined
problem areas. But when expert systems developers tried to
scale up their systems to tackle bigger problems, it become clear
that neither approach was going to work. The problems arose
from simple practicality. There are some people who are willing and able to learn the skills needed to build an expert system,
but there aren't nearly enough for this to be a viable approach in
most fields. If expert systems were going to be used on a wide
scale, it was preferable to have them developed by specialized
expert systems developers who acquired knowledge for each
new field from human experts and whatever other sources were
available.

However, the problem with the interview approach was that
interviews were missing too much. They are fine for some purposes, and they seem easy to use. Most people think of them as
the obvious and most sensible way to gather information. But
in fact, the word *interview* can mean a lot of things, all of them
with limitations. The classic distinction is between structured
and unstructured interviews. In a structured interview, the interviewer has a list of prepared questions, and a flowchart of
follow-up questions is triggered if the interviewee gives a particular answer. It all looks and sounds scientific, but the result
is dependent on having the *right* list of questions, phrased in

the *right* way, with the *right* options available for the follow-ups. By definition, when you're gathering knowledge for a new expert system, you can't know what the right questions are, or the right phrasings, or the right options. If you know enough to design a structured interview, you probably already have all the knowledge you need to build the expert system: a classic catch-22.

At the other end of the scale, there's the unstructured interview. This has been satirically summed up by a cynical software developer: "Okay, tell me everything you know about Ford fuel injection systems."

These problems are well known in fields like psychology, which deal with extracting knowledge and beliefs from human beings. So it was no accident that while at Nottingham I worked in the department of psychology, within a research group specializing in artificial intelligence.

Card Sorts and Laddering

If you interview experts about their specialist areas, sooner or later they'll mention something they have never told you about before. Once, when I was collecting data at Nottingham, I asked a geologist how he could identify a particular type of rock in the field. The geologist went into great detail about rock identification. At one point, speaking about one particular rock, he listed the features he had mentioned for all the previous rocks—like grain size and color—and then said that if you broke a piece of this rock open, it smelled of sulfur. It was the first time he had mentioned the *smell* of a rock as an identifying feature. The implication was that if I wanted to make sure I didn't have any gaps in the information collected, I would need to go back through the list of all the other rocks that he'd mentioned and ask him what they smelled like. That's just one of the problems you regularly run into with interviews: if you're trying to gather information in a way that gives you systematic and complete coverage, interviews aren't a good format. Missing the occasional point isn't a big issue if you're interviewing someone on a television talk show. It can be a very, very big issue if you're trying to gather a complete and correct body of knowledge for a medical expert system, where patients will die if you've missed a key point.

While at Nottingham, I worked for a team of researchers—Nigel Shadbolt, Mike Burton, and Han Reichgelt—who were looking for techniques that yielded information of better quality than that produced by interviews. We were comparing methods systematically, so researchers and developers would have some solid evidence to guide them in their choice of methods when gathering knowledge for expert systems. We started with a long list of candidate methods, soon whittled down to four. One of them had to be interviews; we needed this as a baseline, something against which all other methods could be compared. The other three methods were very diverse.

One of those methods is known as the think-aloud technique. This method is just what it sounds like: you ask the expert to do a task—for instance, you ask a geologist to identify a rock you've never shown them before—and you ask her to think aloud while she's doing it, so that you hear what her thought processes are. It's a widely used technique: a lot of instructors who teach police officers how to drive vehicles get the trainees to think aloud during the lessons. This technique is great for catching the steps that experts don't notice they're taking. However, as soon as you start trying to write software code based on output from this technique, you immediately discover a big drawback. Most people sound like gibbering idiots when they're thinking aloud, because they're alternating between doing the task and telling you what they're doing. The result is that you get semicoherent fragments of their chain of thought. A further problem is that experts often refer to *this bit* or *that color* in a way that's meaningless if you're working off an audio recording. Even if you have a video of the session, you still need to find out whether there's a technical name for the part or color in question. The strengths of thinking aloud are balanced out by its limitations.

The third technique we used is called card sorts. The core concept is that you give the expert a set of objects from his area of expertise, and ask him to sort those things into categories. The objects may be physical things like rocks, or photographs, written names, or written scenarios—the method is flexible. We used a variety of this technique where you ask the expert to sort the things into whatever categories she wants and you let her re-sort as often as she wants, using a different category each time.

The method is excellent for finding out how experts think about their field, specifically how they categorize objects, which can be different from how novices and nonexperts would categorize the same field. This method also gives you systematic information about the categorization: if a geologist is using the category "what it smells like when broken" while sorting a set of rocks, you will be able to see which category each rock fits into. As usual, there is a downside: experts tend to use categories that are technical or subjective, so the interviewer has no idea what the category name means unless the interviewer employs a new technique.

Suppose a geologist sorts some rocks into an "ultrabasic" cat-egory, and you're left asking yourself, "What does that term mean in relation to a rock?" To find out, you need to break out the fourth technique in our set: laddering. Much of human knowledge is structured in hierarchies—the classic tree struc-ture, where a trunk divides into a few branches, and each of those divides into twigs, and each of those in turn divides into leaves. If you ask an expert to explain what a Solutrean blade is, you usually get an answer such as, "It's a long, bifacially retouched flake tool." Next, you ask the expert to explain fur-ther: "What counts as 'long' in this context?" Or "What does 'bifacially retouched' mean?" You'd need to go through several layers of "branches" and "twigs" before you reach the "leaf" level, but you'll get there in the end. It's a powerful, flexible technique that I found invaluable.

Over the next three years, we used these methods to elicit knowledge from experts in fields ranging from medical diag-nosis to geology to archaeology, among others. We crunched the data from different perspectives: how much information we extracted per minute with each technique, how long it took to process the information from each technique before it could be used in an expert system, how many expert system rules you could extract per minute with each technique, and so on. We also tested variants of some techniques—for in-stance, testing whether you get different results depending on whether the expert is sorting physical items, or pictures of those items, or cards with the written names of those items. Ultimately, we had a lot of answers and a bunch of landmark research papers. But, as usual in research, those answers had also brought up a lot of questions, some of which would nag at me for years to come.

Pattern Matching: What You Can Do That a Computer Can't

The Nottingham work was focused on quantifying each method's effectiveness. That was useful, but we didn't have time to investigate more than the numbers. I couldn't help thinking about that geologist doing the think-aloud session, explaining how he would identify the rock sample I had just given him. After half an hour, he was still going strong about the same rock sample: an ordinary-looking rock about the size of my fist. It wasn't even one that he was excited about. The really interesting thing was what happened when I showed him *another* rock sample and asked him a subtly different question. With the first rock, I had asked him to tell me *how he could tell* what type of rock it was. With the second rock, I asked him just to tell me *what* type of rock it was. He identified it immediately, before I had even set it down on the table in front of him. There was no way that he was processing half an hour's worth of knowledge about rocks during that one second: he was identifying it *using a completely different mental route from the one he used in the first session*. But what was that route, and how did it work, and what were the implications?

Those kinds of questions take us back to a problem at the heart of artificial intelligence research, one that is still a big, unresolved problem after half a century: pattern matching. That's something humans can do very, very swiftly. Humans evolved to make snap decisions, because carefully weighing every option is a good way to get eaten by something bigger than you. It's a classic example of something so familiar that nobody notices it anymore. It's also an example of how something so mundane that nobody pays it much attention can be crucial, precisely because it's everywhere and involved in everything.

Pattern matching is all around us. It's the intelligence behind the eyes when we look at the world. But computers simply can't pattern-match the way humans can. A computer will recognize a chair in one photo but not the same chair photographed at a different angle. It can't pick out a park bench in a photo of a pretty park setting. Computers can steer a spacecraft to Mars, but they can't tell you whether a photo shows a chair or a zebra.

Computers use explicit, logical, step-by-step reasoning. That's great for a lot of problems, particularly because humans are not good at this type of reasoning. We can do it, but our brains aren't wired in a way that makes it easy for us, so it's great to have computers that can tackle it for us. Humans are wired for a completely different way of processing information, called parallel processing.

Here's an example that shows the difference between the two approaches. If you're writing bank loan software, you'd build into the system that IF the applicant is under twenty-one years old AND the applicant has less than one thousand dollars in the bank THEN do not loan more than five hundred dollars. The software will work through each of these points in turn, checking whether or not each one is relevant to the applicant, and will then make the decision.

Theoretically, a computer could be wired so that it checks both the IF question and the AND question simultaneously. That's parallel processing: performing two or more tasks at the same time. This is obviously faster than having one person work sequentially through each point in turn, especially if there are hundreds of points involved; it can be massively faster and make things feasible that would otherwise be out of the question. However, it comes at a price. One is that it needs to be managed carefully. The design for hardware and software that handle parallel processing is very different from the design for standard hardware and software.

There are other, more profound differences between one-step-at-a-time sequential processing and parallel processing. Humans, for instance, can glance at a photo of a spotted animal in the snow. They see the spots, they see that it's a dog, they see black-and-white coloring, and they immediately think *dalmatian,* which is the correct answer. In most cases they won't even need to study the image for very long. The brain parallel-processes all the stimuli at the same time and spits the answer out in less than a second.

A computer wouldn't perform anywhere near so well. Right at the start, it would have trouble just working out whether it was seeing spots rather than stripes or random splashes of color. It would have further difficulties working out where the animal ended and the snow began; it might not even realize that it was seeing an image of an animal. Even if it did notice the animal, it would have real problems working out whether it was seeing a dog rather than a cat or a wolf or a zebra. That doesn't inspire much confidence about whether it would arrive at the correct answer.

Pattern matching can be an elegant and efficient problem-solving tool. Doctors harness that power when they inspect a patient's rash. Their eyes take in visual information—the size of the rash, the coloration, how raised it is from the surface of the skin—and their brains swiftly recall all the times they have seen such a thing before. And the doctor proceeds to make a diagnosis. A researcher working with mounds of visual data—charts, graphs, tables—can sift through them and tell at a glance which data is important and which is peripheral. In my own work, I developed a renewed respect for pattern matching when

I realized how its application could profoundly aid human comprehension of visual data. We'll have more to say about this later, because pattern matching plays a critical role in the larger Voynich Manuscript story.

The full story about the human-computer divide is a lot more complex, but that's the basic underlying message. Computers are good at step-by-step reasoning. The wiring of the human brain is good at something very different: a huge number of brain cells handle information simultaneously, with different cells or groups of cells each handling one part of the task. It's an effective way to process the sort of information that is everywhere in the real world: untidy, messy, incomplete information, where part of one object is obscured behind another object, you're seeing the second object from an angle that's completely new to you, and you don't have much time to decide what you're seeing because you're trying to cross a busy street and don't want to end up as a traffic statistic.

But these pros and cons come at a price: you're more likely to commit errors if you're using step-by-step reasoning. You're also more likely to make a swift, efficient error when you misidentify a pattern. That's a theme that was to recur repeatedly when I was trying to make sense of human error.

The Knowledge Within

O NE OF THE MOST UNSETTLING FEELINGS YOU get in research is the sense that you've missed something. I was pretty sure that we hadn't made any significant active mistakes in the Nottingham work—it had stood up well to peer review, and we were using solid methods. I had a nagging feeling, though, that there was something we had missed. I didn't know what it was, but I suspected that it would seem blindingly obvious once I had spotted it. That gut feeling turned out to be right. The key insight came a couple of years later, when I was working on requirements engineering.

The Nottingham work had been focused on situations where experts get it right. My next type of research, at City University of London, was to take me into situations where experts get it wrong—dead wrong. Requirements engineering, like knowledge acquisition, was a new field when I moved into it. Like knowledge acquisition, it had arisen because the old methods weren't good enough for the demands of new technologies. If you're building a spreadsheet for a client and there's a bug in the software, the bug might cause problems for the client and just possibly put them out of business if it leads to a disastrously wrong financial decision, but it won't kill anyone. If you're building safety-critical software that will control the flight systems in an aircraft and there's a bug in the software, it could kill hundreds of people within minutes. The old methods weren't good enough at finding out the key requirements, so people working in this field were looking for solutions in other fields that had tackled similar issues.

This chapter is the story of some adventures in that realm. We'll learn about some of the ways the human mind hides our expertise from us—even when withholding that knowledge could kill us—and some of the specialist methods that researchers can use to tease out that knowledge. I'll show you how a colleague and I were able to spell out exactly how you can get at the essential knowledge in the mind of an expert with the help of a toolkit that anyone in academia or industry could use. That would later turn out to be a key feature of Verifier: all the tools in its toolkit could be used on their own. You don't need to learn all of them to do something useful.

The high school version of science usually tells you that science proceeds by a better new solution replacing a worse old one. There's an element of truth in that, but the reality is more complex. Often the old approach isn't wrong, or even worse than the new one. Often the old approach is only dealing with one part of the problem and is failing to tackle a more significant aspect of the problem. The study of errors is a classic example.

Researchers in formal logic and philosophy have done a lot of work on errors in human reasoning. That literature goes back well over two thousand years. It contains extensive classifications of types of faulty reasoning, from the viewpoint of the internal consistency (or otherwise) of the reasoning. A classic example is the faulty reasoning "Dogs have four legs; that animal has four legs; therefore that animal is a dog."

However, that approach is not the only way of looking at human error. When psychologists like the British professor Jim Reason started investigating human error, they identified types of error that didn't show up in the logicians' and philosophers' classifications. An everyday example is known as frequency capture errors: if you have a front-door key that you use a couple of times a day and an office key that you use dozens of times a day, you'll probably sometimes find yourself trying to open the front door with the office key but hardly ever find yourself trying to open the office door with the front-door key. In situations where you have a choice, if you make a mistake, it will probably be in the direction of doing whatever you do most often. In this case, you use the office key more often than the front-door key, so when you arrive at home, you naturally reach for the office key.

Knowing just this one fact about how humans make mistakes means that aircraft designers will make sure that dangerous options (such as dumping all the fuel from your aircraft) can't be confused with anything similar in a frequency capture error. Thus they design the fuel-dumping controls in such a way that they can't be mistaken for anything else. They also make sure that they take this into account when designing controls for different aircraft, so the habits you've acquired with one don't lead to problems when you port them across to another. At least one fatal air crash was caused by precisely this problem.

As my colleague Neil Maiden and I worked through the literature and worked with real safety-critical systems, we kept seeing the same pattern over and over again. Experts weren't producing a complete and correct account of what they wanted, even when their lives depended on it. One colorful example leaps to mind. It's the story of how some experts—in this case the captains of massive ocean freighters—omitted some key information from their interviews with researchers that could have easily gotten the captains killed.

What's Not Worth Mentioning Can Kill You

One of the biggest changes in large-scale transport over the last twenty years has been the huge growth in container shipping. Containers are a brilliant way to transport items, whether loaded directly onto a truck for road transport or stacked high on a massive ship, like enormous building blocks. But they're not the only way of transporting goods and materials.

Bulk carriers are ships designed to carry bulk cargo—commodities like oil or ore or coal or grain, which aren't well suited to containerization and where there are advantages in transporting them in huge quantities to achieve economies of scale. By the 1980s, bulk carrier ships were enormous, carrying a huge proportion of the world's key commodities. Without them, the global economy would grind to a halt in days. There was one big problem with them, though. They had a nasty habit of occasionally sinking without warning, so fast that they usually weren't able to get off a Mayday message, let alone say why they were sinking. It was costing lives, and nobody knew why. The obvious explanation—that they had run into a storm and capsized—could explain a few cases but didn't work for most of them. There were other explanations that looked plausible, and one that we were working on involved hull stress.

In the old days, metal ships were held together by rivets, which were solid but slow and expensive to construct. Modern ships use welds instead, and although welds are fine most of the time, they can encounter problems. During the Arctic convoys of World War II, there were cases where everyone aboard a ship would hear an explosion, and then the vessel would sink instantly. Survivors and investigators assumed that the ship had been torpedoed. This was one of the acknowledged hazards of Arctic convoys, and one that captains and crews were always expecting. After the war, when naval historians started piecing together the records from both sides of the conflict, they found cases where the ships *couldn't* have been hit by torpedoes because there were no submarines operating in the area. They discovered instead that in the icy waters of the Arctic, ships under stress could crack along the line of a bad weld. The crack spread fast—going around the entire ship at supersonic speed and creating the explosive sound that had been mistaken for a torpedo strike.

That problem didn't only occur in the Arctic. Long after the war, there was a similar case involving a bulk carrier cracking in two and sinking while being loaded in the harbor. When you're loading a bulk carrier, you have to think carefully about how much cargo you're going to put in which section of the ship's hold. If you get it wrong, you can have thousands of tons of cargo at the front and the back of the ship, weighing it down, while the buoyancy of the empty center section is pushing the ship up. It's like having a heavy adult sit on each end of a seesaw designed for children; the likely result is a loud noise and a big break. That's exactly what happened to this bulk carrier: the stresses from the load were higher than the hull could sustain, so the ship suddenly broke in two and sank at its moorings.

If a bulk carrier could sink like that in the calm of a harbor, then it was easy to imagine the same thing happening under a sudden stress at sea. If the hull was already near the danger point when the ship set out to sea, it wouldn't take much to push the stress levels up to the breaking point. Even a small squall, which other ships could ride out easily, might be enough.

Some interested parties—insurers, shipping companies, and others—were working with our team to develop a software system that could monitor stresses on the hulls of bulk carriers, to check whether the hull was reaching dangerous stress levels. The system would include software to monitor the readings from various sensors around the hull. Getting the software requirements right would be a big issue, and thus a juicy case study for our research.

My standard way of starting research in a new area of expertise was to observe the experts in action for a while. You will almost certainly find something that you weren't expecting. After hearing me go on at length about how effective simple observation could be, our collaborators agreed to go down to the seaside and watch our client's bulk carrier being loaded. Until then, they had apparently been planning to design the software based on specifications given to them by third-party specialists.

The observers arrived at the harbor and began watching the loading, which was done via a conveyor belt system, with the iron ore or coal running along and then dropping into one of the holds. All along, the software developers had been making assumptions about the loading process. They had assumed that the loading would proceed at a constant rate, so it would be easy to predict the stresses by extrapolating the line of the stress levels from previous readings. When they asked the ship's captain about this on-site, the man laughed at this notion, pointing out that if it was getting close to lunchtime or the end of the day and there was still a lot of cargo to load, the person in charge was likely to start loading a lot faster. You couldn't easily predict when the stresses would hit dangerous levels, and with a dense cargo like iron ore, levels could change from safe to dangerous very swiftly.

The developers had also assumed that if the stress levels were getting too high, the captain could simply call the person doing the loading and tell them to stop. This provoked more amusement. The captain told them that in some ports there was no communication link between ship's bridge and the loader. Crew members sometimes lobbed objects at the cab of the man doing the loading, since there wasn't any other way of getting his attention and making him stop.

By now, it was clear that another of the team's assumptions had run aground: having the software put out a warning sound if stress levels got too high. Imagine the sound level when somebody drops tons of iron ore into a huge metal ship from forty feet up: it would need to be an impressive warning sound to be heard. Having flashing lights signal a warning also wouldn't work. The captain said there was so much glare on the ship's bridge when loading cargo in a sunny locale such as Australia that he sometimes had to put cardboard boxes over display screens on the ship's bridge, with eyeholes cut in them, just so the crewman could see the screen despite the glare.

Clearly, the previous team had missed some key points when they interviewed the experts in a sterile office environment. Those key points had been easily detected using *observation*. The captains hadn't thought to mention this stuff because they hadn't been asked about it. They thought these details were obvious and barely worth mentioning.

If you changed the methods you used to gather information from experts, you would increase the quality of the information you got out and catch observations that would have been missed otherwise. So the problem wasn't about the experts' willingness to cooperate, or about their level of expertise. This was beginning to look like a puzzle, but I still couldn't work out what the picture would be.

And then, one day, Neil and I realized what the picture was.

ACRE

Neil and I wanted to get a paper into a conference at Keele University. There was one minor problem: the deadline for submitting articles was two weeks away, and we hadn't even begun to think about what we would write, let alone start writing it. We scheduled some time to kick ideas around over a cup of coffee, and were soon talking about what other people were missing in their research.

We were wrestling with the same familiar problem of why experts wouldn't give you a complete and correct list of their requirements—as if there was something intentional about it. And suddenly it hit me: everyone might have been asking the wrong question all along. Everyone had been asking why people *wouldn't* give you that complete and correct list. What if we asked a different question instead: what if we asked why people *couldn't* give a complete and correct list? In other words, what if they were incapable of sharing everything they knew because they simply could not access that information?

That question changed everything. To anyone with a background in psychology, the answer was obvious: there are different types of memory, skill, and knowledge, each stored and processed in its own particular way within the brain. When you started from that direction, it was ludicrous to think that people might be able simply to pull out all the relevant information during an interview or questionnaire. Memory just doesn't work like that. People with experience in their field build up a lot of knowledge that they're unable to put into words. It's there, it's usually accurate (remember the rule about experts knowing more than novices), but it's often something they use, not something they can say. It's called by a lot of names, such as "gut feel."

To put it in terms of our lessons about experts, you could think of it this way:

LESSON 6
Experts often don't know what they know.

The implications of this insight were far-reaching. If you knew even the basics of how the brain handles different types of memory, skill, and knowledge, and you knew a reasonable range of ways of eliciting information from people, then you could easily put together a menu that mapped techniques onto types of knowledge: "For *this* type of knowledge you have to use *this* technique for information gathering, but for *that* type of knowledge you need to use either *that* technique or *this other* one."

There was a lot of literature about these different types of memory. But no one had thought to assemble them into a practical framework so that a researcher could just work his way down the list, choosing the right technique for the type of memory he was trying to elicit from the expert. Neil and I did a rough outline sketch of the paper, divided up the work, wrote our respective sections, and bolted them together. The result was something that we called the ACRE framework—standing for ACquisition of REquirements.

At its core, this says: here's how you can divide knowledge, memories, and skills into types and subtypes. Here are the implications that go with each of these types and subtypes— implications for how you gather information from people, what to watch out for, which methods to use and avoid, and why. If you're wondering why your customers appear to be unable to make decisions, there's an answer in the framework. If you're wondering what to do about that problem, there's an answer to that question, too. And it's all based on evidence, often huge amounts of evidence, from the relevant areas of expertise in psychology, neurophysiology, and elsewhere.

Previously, there were two main approaches to choosing the right method to get information out of people. One was to choose the method you liked: each method had its advocates arguing with the advocates of the other methods. The other was triangulation, which boils down to saying that if you got the same result from two or three different methods, that result was probably in the right ballpark. That might seem sensible, but how could you tell if you had happened to choose two or three methods that all had the same weak spot, and which all happened to give the same wrong answer?

Here's the framework. It has grown and changed slightly from the original one that we presented at the Keele conference, but it's the same at heart. It's divided into four main categories of knowledge, memory, skill, and so on—I'll refer to them all as "knowledge" for simplicity through the rest of this description.

First Category: Future Requirements

Imagine that you're looking for a memorable vacation, and the travel agent asks whether you'd like to book a trip to see the Redentore. Would you love it or hate it? Most people wouldn't have a clue, because they have never heard of it. The travel agent explains that it's a big festival in Venice. Now you can start to have an opinion, but there are still questions you will probably want to ask. Think back to the people using the traditional waterfall model of developing a new system, or to the traditional questionnaire with predefined answers. How would they cope with this issue? They'd ask the question, get a noncommittal response, and move on. They would not consider it part of their job to educate the expert, consumer, or client.

The more novel a product or concept, the more it runs into this problem. The solution is simple: the interviewer must allow for several iterations of questions and be prepared for dramatic changes in people's responses as they find out more. If possible, you must show your participants photos, video clips, and mock-ups, so that you're not filtering the information through words and leaving out something that will be significant to the client, even if it doesn't seem significant to you. For instance, the Redentore features a huge fireworks display at its climax—pretty significant to clients who either love fireworks or hate them, and not a standard feature of all festivals.

People are good at spotting things known as "affordances"—ways you can use a product that may never have crossed the designer's mind. I was at a conference once where someone described the concept of a motorized hospital bed that could be programmed to move on its own to a chosen destination. The inventor had focused on how helpful that motorized bed would be if people used it properly, but the affordances that the bed offered to pranksters were frankly terrifying.

Explicit Knowledge:
Memories in Search of Lost Time

Most people think memories are all handled by the brain in much the same way. But in fact, there are different types of memory. In the ACRE framework and its later refinements, we distinguish between various types of memory. To give you an idea of how we tackle them, here's a brief overview of one—episodic memory. This type of memory is about episodes—experiences you had at a particular point in your life, as opposed to *semantic* memory, which deals with "timeless" facts, such as the capital of France or how many yards there are in a mile.

One of the most striking forms of episodic memory is known as flashbulb memory. It's the "What were you doing when you heard about 9/11?" type of situation, where people often have vivid memories of exactly where they were and what they were doing. Those memories looked like a brilliant way to gain insights into how memory works. So some researchers, notably the American psychologist Elizabeth Loftus, began investigating just how the human brain could remember some moments with such photographic detail. They discovered that actually the brain didn't remember them that well at all.

A classic example is the German-born American researcher Ulric Neisser, who had a vivid flashbulb memory of hearing the radio commentary for a baseball game being interrupted with breaking news of the Japanese attack on Pearl Harbor. But Pearl Harbor took place on December 7, not during baseball season. So if this researcher's vivid, detailed flashbulb memory was wrong, how many others were, too?

Over the years, Loftus and her colleagues, including Alan Baddeley, pieced together a compelling answer: flashbulb memories weren't trustworthy, regardless of how detailed they were and regardless of how fervently the person believed that the memory was accurate. This finding had devastating implications for questions such as how much we can trust eyewitness testimony in courtroom trials.

Another key finding was that memory isn't like a photograph, which may become faded but is basically a faithful record. Instead, it's more like an artist's sketch, where the artist makes decisions about what to include and may misinterpret a feature of the scene.

This process applies not only when the person retrieves a memory but also when he encodes it, or saves it to the not-so-hard drive of his brain. A classic study involved showing participants images of one person threatening another with a knife. The participants were good at remembering that there were two people involved and one was threatening the other with a knife, but they weren't as good at remembering which one was holding it. Again, the implications for eyewitness testimony were disturbing.

Explicit knowledge may therefore be wrong or distorted, but at least the person you're interviewing can readily access it. The situation is different with the next main type of knowledge, which we called semitacit knowledge.

Semitacit Knowledge

Semitacit knowledge can be accessed via some techniques but not all. The story of the bulk carriers is good example of semitacit knowledge. The captains had the information and they could articulate it, but as we saw, interviews were not a good way to get that information out of them, because they wouldn't volunteer it, and researchers would not know to ask for it. Only on-site observation could reveal it.

Another good example of semitacit knowledge is short-term memory. Long-term memory is what it sounds like: it's what we use to store memories for any significant length of time. This can vary between minutes and decades; there are plenty of centenarians who have vivid, accurate memories of their childhood a century ago. There's a decay rate with long-term memory, however: you'll lose most of today's new memories overnight, and you'll continue to lose some of today's memories over the next couple of weeks or so, but any memories that make it past the two-week point will probably stay in your brain for the indefinite future.

Long-term memory not only provides long-term storage; it also has a huge capacity and can easily hold a century's worth of memories. Short-term memory is different. It's what we use for remembering a phone number until we can write it down: a short-term mental scratchpad. It has a tiny capacity—about seven items—and a tiny duration: a few seconds, unless you keep repeating the phone number to yourself so you don't forget. It's incredibly useful, and yet people never really notice it.

Until you're doing something like designing the software interface for aircraft pilots. Then you suddenly realize that it's very, very useful to know which information the pilot needs for each task. You discover that a lot of the time the pilots are checking a number on a dial, and they're holding it in short-term memory for a couple of seconds until they don't need it anymore. If you design the new system so that they can't easily get at that number when they need it, you could be facing big problems. The trouble is that because the contents of short-term memory are lost within seconds, no amount of interviewing after the flight will get at that information. The pilot's memories about those numbers might be accurate, but would you feel inclined to gamble your career and two hundred lives on that possibility? What you need to do is to get at the contents of the pilot's short-term memory during those ephemeral seconds. One way to go about that is direct observation: you watch closely what the pilot is doing, while taking care not to interfere with how he would normally operate. Another is to use the think-aloud technique.

Short-term memory crops up everywhere. It's a big issue in software interface design, where the visitor to your website typically decides within two seconds whether to bounce on from your site to another. You need to make that interface as user-friendly as possible, and for that you need to know just what is going through people's minds when they try to use it.

There are several other forms of semitacit knowledge. One that I encountered repeatedly in requirements engineering is taken-for-granted knowledge. One of the basic principles of communication, described by the British linguistic philosopher Paul Grice, is that people don't mention in conversation anything that they can safely take for granted that the other person already knows. You don't say, for instance, "My aunt, who is a woman," because you can safely assume that anyone who knows what the word *aunt* means will know that an aunt is a woman.

As we saw with the sea captains, this principle doesn't work as well when you're an expert. It's horribly easy to assume that just because you and the people you routinely work with are familiar with a concept, you can take it for granted that everyone else is as well. I struggled with this issue for the first time when I was dealing with the geologists at Nottingham and realized that while *they* knew if a rock was fine, medium, or coarse grained, I never would unless I got them to explain those terms.

So that's a brief description of semitacit knowledge. It's tricky stuff, but it's knowledge that you *can* get people to talk about if you know how to prompt them. That leads us to the most elusive, most mysterious of all human knowledge: the stuff you don't know that you know.

Tacit Knowledge

Different fields use the term *tacit knowledge* to mean different things, and they are similar enough to cause confusion and different enough to cause problems. We used the term in the strictest sense, to mean knowledge that people couldn't accurately verbalize even if they wanted to.

Pattern matching is an example of tacit knowledge in the strict sense. You might know what a particular face looks like— Bill Clinton's, for instance—but trying to describe a face unambiguously in words, to a level where nobody reading your description could mistake it for any other face, would be a waste of time. Words are simply the wrong form of representation for that type of knowledge. No wonder so many novelists describe a character as "ruggedly handsome" or "astonishingly beautiful," and march on to the rest of the plot.

You'll encounter another type of tacit knowledge if you ever ask someone to think aloud while driving. When we first learn to drive, we think through every step and accept directions from our instructors, who articulate what we need to do every step of the way. But as we become more expert, starting a car, shifting gears, and so on become so automatic that we have a hard time putting that experience into words. We call these compiled skills. As a rough rule of thumb, if a person can carry out a nonverbal task while holding a sensible conversation with you, they're probably using a compiled skill to carry out that task.

For example, an expert driver will typically be able to give you a sensible commentary most of the time she's driving but will stop talking when she needs to concentrate for an intersection. What's happening is that she has practiced some driving skills to the point where the brain has set aside pathways for that activity, so the activity doesn't need conscious attention anymore. She can comment on the overall strategy of what she's doing—slowing down slightly because she can see a novice driver ahead, for instance—but she's no longer conscious of the low-level actions she's performing when she implements a strategy. Because those neural pathways for shifting gears and so forth are nonverbal, the person can't put into words what low-level actions she's doing, even if she would like to.

In fact, if she tries to figure out what she's doing *while* she's doing it, the figuring out will probably get in the way of the activity and make everything come undone. Skilled sports coaches know this, even if they're not aware of the underlying neurophysiology; the best way to check whether a downhill skier is actively shutting out external distractions is to ski beside him and tell him just how well he's doing. When an athlete talks about "being in the flow," he is talking about that blissful moment when nonverbal compiled skill takes over, allowing him to perform with no conscious thought involved, through muscle memory alone.

Compiled skills are efficient, much faster than the verbal thought version. That's why driving behind a student driver can be so frustrating: they're having to think through every action.

However, compiled skills are also prone to "strong but wrong" frequency-capture errors: you see a familiar situation, and your compiled skills react the same way as usual, even if that's not the right way this time. You act without thinking and dump your aircraft's fuel in the ocean because you've confused the instrument layout from your previous plane with the one you're flying now. Or you take out the office key when you arrive at your home. Experts are just as prone to this kind of miscued error as the rest of us.

Sometimes tacit knowledge is not taught by carefully articulated steps, the way driving is, but rather learned by doing, almost through a process of osmosis. We call this process implicit learning. If you've asked a baker how she knows just how much a pinch of salt is, or a brewer how he knows just when to add the next batch of hops to the mixer, you will encounter one of the stubborn traits of an implicitly learned task: the expert has no idea why he does what he does; he just chalks it up to a kind of hunch.

So, how can a researcher find out what's really going on? By definition, you can't access tacit knowledge via questions. You can *observe,* though, and see what patterns of behavior are going on. Once you've got an idea about those patterns, you can experiment—modify the context, and see what happens when the person tries performing the skill. It may turn out that the expert brewer is actually responding to a change in the *sound* of the mixer laboring against the mash, and that's how he knows when to toss in the rest of his ingredients. He has been doing this for years and doesn't know that he has been cued by sound. He performs his task when he thinks the time is right, and can't explain his actions other than saying he just knows in his gut that the time is now.

You can see how tacit knowledge might affect science, industry, medicine, and other fields. Experts may be very, very good at what they do, but every day they may make a tacit knowledge skill error that no one will notice. There's a mass of research showing clearly that experts are as liable to some types of misjudgment as the rest of us. Often, they drastically underestimate the significance of a few low-level factors that they have taken for granted and overlooked.

The error of a businessman or a stockbroker or a physician might boil down to a quick poorly made decision. Such an error is simply not as visible as a pilot dumping fuel out of the aircraft. But in an era of financial crises, a stockbroker's decisions could have knock-on effects that change even more lives than the pilot's error.

3

The Imperfect
Expert

THE DAY NEIL AND I WERE GOING TO BE PRESENTING our new framework for getting knowledge out of experts, we had two problems. One was that I had picked up a bug and almost lost my voice. It was okay as long as I kept speaking, but if I stopped, there was no guarantee that my voice would start up again. The other problem was that, while one of the earlier speakers was giving a talk, we'd been studying the body language of a couple of big players from our industry who were sitting in the audience. Even the rawest beginner can spot a signal like two audience members lifting their copies of the *Times* out of their briefcases and starting to read. This was the audience that Neil and I would be pitching to. If we could grab the undivided attention of those *Times* readers, then we were onto something big. After seeing their reaction to the previous talk, though, we were well aware that it was a big *if*.

This is how we started our talk: "Why is it that people won't give a complete and correct list of their requirements in surveys and interviews—even if their lives literally depend on it?" The two newspaper readers looked up. It was the reaction we had been hoping for. When we started describing how we had pulled together an assortment of literatures into a neat, coherent framework, they stowed their newspapers, and we knew we were home free. My voice held out just long enough to reach Neil's part of the talk. We spoke for twenty minutes, and the questions at the end made it clear that we had gotten people very interested. Afterward, one of the most senior researchers in the field came up to congratulate us. It was a massive relief, after all the years of trying to make sense of the elicitation problem— actually getting laymen and experts alike to open up and talk to you—and the usual nagging worry about whether we might just have been kidding ourselves about the soundness of our solution, especially when it had been put together so quickly. My memories of the rest of the conference are a warm, fuzzy blur.

One thing ACRE had going for it was that it made clear, hard predictions about the strengths and weaknesses of each approach—interviews, think-aloud, observation, and so on. Our framework was very different from a vague, meaningless statement that all methods of unpacking knowledge were equally worthy; instead, it was a tough, rigorous approach that worked well with the engineering mind-set of the people who actually build safety-critical systems. We were hoping to move on to new challenges now, using our findings to tackle larger problems. I wanted to see how well a researcher could apply ACRE to problems in market and product research, to find out what the public really wanted. Both of us were still obsessed with human errors. We had a dream of unifying the disparate bodies of scientific literature on human error into a single, coherent framework, a little like what we had just done with ACRE.

But that was going to be tough. The literature on error spelled out some pretty damning things about all human beings, not just experts: Humans have trouble thinking logically. Humans have trouble working with math, specifically probability. Humans have biases. And everyone looks at problems from their own point of view, leading to potentially disastrous communication problems. These were huge problems for any researcher trying to comprehend the expert mind. They were also a nightmare for professionals trying to tackle problems in their fields. Simply put, if you don't know that these pitfalls are waiting to ambush you and you don't work hard to correct for them, you're going to fail again and again. The way Neil and I saw it, human error was going to be a problem for a long time.

In fact, I would find a way of making sense of human error just a few years later, with the invention of the Verifier approach. While I was working in London, though, I didn't think there was much chance of anyone seeing an underlying picture for human error in the way that Neil and I had just done for elicitation methods. To show why I thought that, here's a description of the problem we were grappling with.

Logical Fallacies and Human Error

ACRE was comparatively easy because we only had to fit together a body of research that went back a century or two. By contrast, a lot of the literature about how people make mistakes dates back to ancient times and is still very much in use today.

The classical logicians and their medieval successors identified a huge number of logical errors, such as the one that goes, "Dogs have four legs; that animal has four legs; therefore that animal is a dog." Their categorization is still in routine use by logicians today. It's easy to show that the reasoning about the dog is wrong by simply producing a cat as evidence; there's a more complicated way of showing that it's wrong by using an area of math known as set theory. Either way, it's clearly and obviously flat wrong.

However, that particular type of faulty reasoning is still all around us. A few minutes watching a political debate or scrolling through the comments section of any blog, online news article, or Internet forum will turn up classic examples of logical fallacies that were first identified by medieval scholars and their Roman predecessors. Threatening others with force or loss of their jobs for objecting to an argument (*argumentum ad baculum*); attacking a speaker's argument based on some facet of her personality (*argumentum ad hominem*), or attacking someone's reasoning by accusing him of being a hypocrite (*tu quoque*)— they're all dodgy rhetorical devices, and they all involve types of reasoning that were shown to be faulty over twenty centuries ago.

More worryingly, there are some things that formal logic either can't handle or can only handle awkwardly. A classic example of thinking that hits a wall is Zeno's paradox, which goes back almost two and a half thousand years. Zeno, a Greek philosopher who lived in southern Italy, invented several paradoxes to demonstrate the limitations of traditional logic. The one that provides the most scope for entertainment in a lecture on formal logic, especially if you happen to have brought your archery equipment, is the paradox of the arrow. Imagine that you're an archer, shooting at a target. Before your arrow can hit the target, it has to travel halfway there. That step in the argument is reasonable, and you can easily demonstrate that it's true. But before the arrow can travel to the halfway point, it has to reach the quarter-way point. And before it can get there, it has to cover an eighth of the distance. You can see where this argument is going. You can keep dividing that distance an infinite number of times. Thus the arrow will never be able to leave the bow, because there will always be another, smaller distance that it has to cover first.

Obviously there's a huge fault in this reasoning: you can easily demonstrate that arrows can and do travel to their target. But it took logicians over two thousand years to finally nail down the flaws in this reasoning, and they couldn't do it alone. They needed the help of mathematicians, who resolved the Zeno matter via some heavy-duty mathematics that go into territory almost as weird as the original paradox.

This probably doesn't sound surprising. In the modern world of specialists, we accept that an expert from a different field can come in and use tools that are different from yours and get some leverage on a logjam. But if you don't know you're in the grip of a fallacy, you won't think to invite someone else to help you.

Everyone involved in Zeno's paradox, including Zeno himself, was aware from the outset that there's a flaw in the reasoning, because it leads to an outcome that is so obviously wrong. But what if it *didn't?* What if it led to an outcome that looked perfectly reasonable? Would anyone have spotted the flaw in the reasoning then? Probably not.

That raises the unsettling question of how many key assumptions at the heart of modern research might be as faulty as the assumptions at the heart of Zeno's paradox. But these assumptions go undetected because they don't produce such visibly wrong conclusions. More subtly, many of the key assumptions in a field might be true only for a limited set of conditions.

Someone who does science or works in industry might well say, "Well, *I'd* never make a mistake like that." But physicists and astronomers are well aware of this issue, because they can hardly pass a day without bumping into the difference between Newton's physics and Einstein's physics. Einstein didn't show that Newton was wrong; instead he showed that Newton's physics were a correct but limited subset of the laws needed to describe motion. It's only when you start moving beyond the familiar conditions on this planet that the limitations of Newton's work start to become apparent—for instance, satellite navigation systems need to factor in Einstein's findings to remain accurate. If you use something outside the conditions it was designed for, then you run into problems; that's as true for physics as for domestic appliances. So you need to know about the limitations of the methods you're using, and you need to know which types of expertise to bring in.

This issue of appropriate expertise has far-reaching and tangible implications. Imagine that you're designing an air traffic control system. You want the control panels to be laid out so there's minimum risk of the operators making a mistake. If you asked an expert in software interface design for advice about this, one of the first things she would tell you is: "Don't have any critical controls differentiated only by color—about 14 percent of men are color-blind, and you don't want them hitting the big red button when they should be hitting the big green button."

There's plenty more good advice for you out there, but you'll have to collect it from different fields of expertise. If you want to know how often someone is likely to transpose a given pair of numbers by mistake, or how many more mistakes someone is likely to make if they've missed a night's sleep, or what proportion of the population has trouble telling the difference between left and right, then you need to go to the right type of expert. But if you don't realize that there are gaps in your expertise, such questions wouldn't occur to you, and you'd never get around to asking a different specialist to help you.

Suppose you're doing a public opinion poll, and you ask an expert logician in the philosophy department of a major university what the likelihood is of someone committing a specific type of logical error known as a conjunction fallacy. It's highly unlikely that the logician would be able to give you a number. That's just not the sort of thing they do. This doesn't mean there's something wrong with how logicians approach the universe; different disciplines focus on different topics. You wouldn't expect your driving instructor to do your car maintenance, or your local mechanic to give you driving lessons. It's the same in the scientific world, where there are different specialties. A key skill is to know which specialist to turn to for possible solutions to a particular problem.

In the case of human error, we've already met two specialist fields that have a contribution to make. Logicians slice up the problem one way. Human factors specialists, like the people who design software interfaces or machine control panels, slice it up in a different way. There are a lot of other ways, too, and they all shed light on the problem of how experts make mistakes.

Linda: The Problem of Calculating This *and* That

Most of the characters in this story are larger than life. Linda is more of a low-key figure. She's quiet, conscientious, and good with numbers; she's fond of animals, eats organic foods, and does meditation. You probably know someone just like her. There are lots of Lindas working as bank clerks or administrators, quietly and efficiently keeping things working; societies like Greenpeace and Friends of the Earth have a lot of Lindas among their members.

Back in the 1970s, some American psychologists made some surprising discoveries about Linda. Those discoveries were particularly surprising because Linda is a fictitious character.

One of the classic texts on human error is a book edited by Daniel Kahneman, Paul Slovic, and Amos Tversky, a trio of scientists who did pioneering work on this subject. Their book was a compilation of studies like the one about Linda, which became an iconic case. There have been numerous variations on that study since then, but the core concept is the same. People are given a short description of Linda like the one you've just read. (I've slightly altered the description so it doesn't look dated.) They are then asked these kinds of questions about her:

How likely is it that Linda is a bank teller?

How likely is it that Linda is a member of Greenpeace?

How likely is it that Linda is *both* a bank teller and a member of Greenpeace?

Most people judge the third probability to be the highest—she's more likely to be a bank teller *and* a member of Greenpeace than to be *just* a bank clerk or *just* a member of Greenpeace.

And they'd be completely wrong.

Logicians call this a conjunction fallacy. The third option in the list above actually represents the area of overlap between the group of people who are bank tellers *and* the group of people who are members of Greenpeace. Any time you have an area of overlap, it *must,* by definition, be smaller than the sum of the two groups that are doing the overlapping. A good way to show this is with a Venn diagram (see figure 2).

If that explanation doesn't get the point across, there are other ways of illustrating it. One is to use a different example. For instance, you could ask someone to estimate how likely it is that Linda was born on a Tuesday. If we ignore minor statistical wobbles, the answer is about one chance in seven: there are seven days in the week, and she's equally likely to have been born on any of them. You could then ask him to estimate how likely it is that Linda was born in March. Answer: about one chance in twelve, because there are twelve months, and she's about equally likely to have been born in any of them. So, what's the likelihood of Linda having been born on a Tuesday in March, as opposed to just any Tuesday or just any day in March? Most people would realize that it's a much lower chance.

Kahneman and his colleagues, however, had no intention of stopping there. For starters, why did people so often make this error about how likely it was that Linda was a bank teller and a

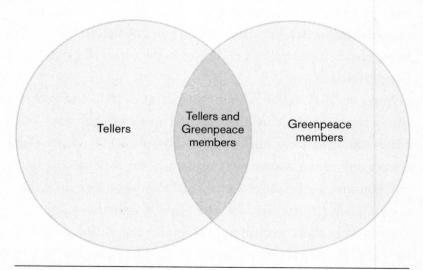

FIGURE 2

member of Greenpeace? Were people just misunderstanding the question or perhaps using a different interpretation from what the researchers intended?

These questions, and others like them, spurred a generation of research in a subdiscipline that became known as "heuristics and biases" within the larger field of judgment and decision making. That research produced some unexpected findings.

Judgment and Decision Making

The name of the heuristics and biases school comes from one of the main findings of Kahneman and his colleagues. People tend to use a lot of heuristics—mental "rules of thumb"—when they're solving problems, rather than thinking the problem through using logical step-by-step algorithmic reasoning. In the real world, you're usually dealing with imperfect, incomplete information, and having to deal with it fast. Experts use a lot of rules of thumb to cut a problem down to a manageable size; everybody uses similar lower-level rules of thumb with everyday tasks. But sometimes those rules of thumb lead us to make errors. They no longer belong to the realm of logic; they become *biases*.

Suppose that you're about to start your familiar morning drive to work. You discover that you can't get into your car because the key won't open it. You probably assume that there's something wrong with the key, or the lock mechanism, and you investigate those. But formal logic would require that you *also* investigate whether someone might have sneaked your vehicle away overnight and replaced it with another of the same model and color, parked in the identical place where you left yours. We generally tend to ignore such outrageous possibilities in order to cut the problem down to a manageable, realistic size.

These rules of thumb work well most of the time, as they did in the possibly switched car, but they don't work *all* the time. If you know what those rules of thumb are, you can start predicting where people using them are likely to go wrong. Researchers using this approach have put together an impressive list of the heuristics and biases that human beings use.

One of the many neat things about this list is the way a lot of it fits snugly into what is known about how the brain works. For instance, people tend to overestimate how often memorable events take place compared to dull ones; that makes sense, because the brain uses multiple cues when it's trying to retrieve a memory from storage, so a vivid memory that links with multiple cues (smells, colors, sounds, and episodic details) has a far better chance of being retrieved than a dull memory with few outstanding "hooks" to improve its chances of being found.

This field of research produced a lot of solid findings, which gave powerful new insights into how people operate and make mistakes. But there were some parts of the story that were still untidy and looked as if they might produce another twist in the plot. That's just what happened, via the work of Gerd Gigerenzer and Peter Ayton. A neat example of their findings involves Gigerenzer's study of students in Turkey trying to predict soccer results in Germany, with surprising outcomes.

The Count: The Problem with Probability

The German soccer results study is a good example of how, if you know what you're doing, you can take something so obvious that nobody questions it, turn it inside out, and discover a whole new world hidden inside. Expecting Turkish students to predict the results of German soccer matches is like asking an American student to predict British FA Cup soccer results. It looks like a pointless exercise, the sort of thing that regularly gets pilloried by the popular press as an example of academics wasting the taxpayers' money.

Except that the students *predicted those results at levels way above chance.* In fact, their predictions were nearly as good as the predictions of German soccer fans and experts. That result made people sit up and take notice. Gigerenzer was well aware that the mental strategies underlying the students' success could be applied to some practical questions, such as assessing which patients to treat first in an overstretched hospital emergency room. As if triage weren't enough, the study could also help researchers understand how humans process information about risk. What's going on with that chain of reasoning? To get a handle on what the research found, you need to face the depressing fact that most people are really, really bad at some types of mental arithmetic.

Humans are pretty good at counting, on the whole, and not too bad with addition and subtraction. We're generally not good at multiplication, though, and most of us are almost useless when it comes to division, unless we have paper and pencil (or preferably a calculator or spreadsheet) to help us. One reason is that doing division places a heavy load on people's short-term memory.

Here's what happens. When you're doing mental arithmetic, you need to hold various numbers in your memory: the number you're dividing, the number that you're dividing it by, the number produced by the first stage of the division, the remainder left over from that stage, and so on, all the way down the line until you have completed the entire division. But since your short-term memory can hold only seven chunks of information, about the length of a telephone number, it doesn't have much chance if you're trying to do math involving more than very small numbers. Yes, you can learn strategies for doing impressive feats of mental arithmetic, but those are strategies that you have to *learn,* precisely because they let you work around the limitations of the standard-issue human memory.

As a result, most people are terrible at any calculation involving complex probabilities. We're often pretty good at identifying which things we need to take into account, but then we tend to misjudge how much allowance we need to make for each. For instance, if you're predicting the outcome of a game between a couple of soccer teams that you know well, you'll want to factor in that one of the players is recovering from a knee injury, another has been going through a bad streak recently, and one team is playing on its home territory. But Gigerenzer's Turkish students didn't have access to any of this information, so they fell back on a simple strategy that can be boiled down to one sentence: "If you've heard of one of the teams in the match, then predict that it will win; if not, just guess." That single rule of thumb turned out to work remarkably well because of a simple principle about how the world works.

Most teams are named after their hometowns. You're more likely to have heard of a big hometown than a small hometown, and big hometowns can afford better players, so their teams are more likely to win. The same applies to teams whose names aren't linked to towns: you hear the names of the successful ones. So it's a pretty effective strategy. The German football experts also knew this strategy, but they weren't taking into account just how good a predictor it was. They were instead attributing far too much weight to factors such as which players had injuries, which actually didn't make much difference to the outcome.

This finding came out of classic studies in the heuristics and biases tradition. What Gigerenzer and his colleagues did with findings like this, however, was to mount a flank attack on the Linda problem, in a way that appeared to clear up one mystery but created a bigger one in the process. In the original Linda problem, researchers had asked people to estimate how likely it was that Linda was a bank teller *and* something else, such as a feminist. There were variations on the question—in some studies, for instance, it was a question about the likelihood of two specified political events happening compared to each event separately. Whatever the topic, the results usually came back the same: people tended to commit the same conjunction fallacy, regardless of their level of intelligence or expertise. The one partial exception was people with specialist knowledge of math or statistics, and even they weren't perfect.

Linda had a lot of practical implications for the real world. In one recent tragedy, a mother was convicted of killing her two babies, largely on the testimony of an expert pediatrician who told the court that the likelihood of two children from the same family dying of Sudden Infant Death Syndrome was so remote that it could be effectively ignored. It turned out that he had fundamentally misunderstood the statistics—his expertise was in pediatrics, and when he strayed out of that field into statistics and probability theory, he made a major blunder that put an innocent woman in jail and wrecked her life. She won her appeal eventually, but died soon after, a tragic figure.

It's not an isolated incident. There are plenty of other cases of medical experts who completely misunderstood statistics, with devastating results. One surgeon performed a large number of needless mastectomies because he had completely misunderstood how to calculate the statistical likelihood that a particular test result meant his patients actually did have breast cancer.

As we said before, we usually see only the highly visible failures. However, exactly the same misunderstandings happen every day, when ordinary people make mundane decisions involving estimates of how likely it is that something will happen. Usually the results are minor, often invisible; for instance, you end up spending more on your mobile phone plan than you might have, because you miscalculated how much you would be using the phone. Sometimes these errors are serious, sometimes fatal, but those are usually normal accidents, not newsworthy except in the local paper for a couple of days, so they go unremarked except by the families and a few close friends.

What researchers like Gigerenzer, Ayton, and their colleague George Wright did was to recast the Linda problem so that it wasn't about estimating probabilities anymore. Instead, they framed it in terms of *frequencies:* "Out of a thousand people matching this description, how many would be a bank teller?" or "Out of a thousand people matching this description, how many would be both a bank teller *and* also a member of Greenpeace?" The precise wording varied, but the underlying core was the same: instead of calculating probabilities as decimals or fractions, they were asking people to do simple counts—to say how many cases there would be within that sample of a thousand people. The results were striking. When you phrased the question in terms of frequency counts—forcing people to think about how many people out of a thousand were likely to satisfy the attributes—the Linda effect melted away. People stopped committing the conjunction fallacy.

Some of these findings could be applied directly to the real world, and that's already happening. Gigerenzer works with emergency staff in hospitals to help them swiftly and efficiently identify the key predictors of patient outcomes. Their input has improved patient survival rates. There's also an increasing trend in medicine to represent possible outcomes in terms laymen can more easily understand. The experts are more often using natural frequencies—"This many patients per hundred thousand will encounter problems" or "This many tests per thousand will produce a false positive" rather than probabilities—"a 0.0003 probability of this outcome but a 0.0027 probability of that one." So the team's work is definitely making a difference. Gigerenzer's book *Gut Feel* is a great introduction to this work—quick-paced, clear, and accessible.

Many questions remain, and as usual, the answers to one set of questions have generated a new batch of unanswered ones. Researchers need to work out *why* people are prone to getting the Linda problem wrong. And it would help to know if other heuristics and biases could be fixed by some apparently simple reformulation of the problem. It's going to be a big challenge, because the probability problem crops up all the time, not just with the general public but also with experts who ought to know better.

Hot Hands

All the ways of estimating probability are based on explicit assumptions. One of the most common assumptions involves things being independent of each other. Suppose, for instance, that you have two completely separate safety valves on a device. The likelihood of each going wrong at a given time is one chance in a thousand. If the two valves really are completely independent of each other, so that what's happening to one will not affect the other, then the likelihood of both valves going wrong at once is calculated by multiplying the two likelihoods together—in this case, giving you a likelihood of one in a million.

That's the assumption that the pediatrics expert was making when he calculated how likely it was that the two babies had both died of SIDS. He was making that calculation based on the assumption that each SIDS death was a completely separate, un-related event—like a dark image of the chances of two children from the same family losing a fatal lottery on separate occasions.

But he hadn't considered that there might be a complicating factor at work. And a lot of complicating factors applied to this case. There were good reasons for suspecting that there was a genetic factor involved, and if one child in a family had the un-derlying problem, there was a higher likelihood that some of the other children would also have it. There was also evidence that an environmental factor might be involved, such as whether the child was put in the crib face down or face up. Again, this would be systematically more likely within the same family. When you put all these factors together, the odds of a second death in the same family changed dramatically. The likelihood was still tiny—parents who have been through that tragedy once are extremely unlikely to have to go through it again. But when you multiplied that tiny number by the number of babies in the country, it became *more likely than not* that some unfortunate family somewhere would have two deaths. But the expert had missed that point. And an apparently arcane statistical assump-tion ended up having very big implications for people's lives.

If you're looking for a deserving cause to champion, there are plenty of probable miscarriages of justice. It's easy to spot possible errors of commission. It's harder to spot things that aren't there, the errors of omission, but they exist too, and they can be bigger killers. There's widespread acceptance now—in medical research, for instance—of doing science not only by conducting your own experiments and clinical trials but by systematically reading all the scientific papers in a particular field and drawing conclusions based on the patterns that emerge. In a sense, the papers, not a carefully designed experiment, provide the evidence.

This approach has been effective in some cases. This is how medical researchers made some completely unexpected discoveries about how statins—cholesterol-lowering drugs—affect the risk of death from heart attacks. Done properly, these "paper studies" can be an invaluable tool. But, like statistics, they depend on some key assumptions. One key assumption is that your review has included all the relevant variables.

If medical researchers a century or two ago had performed a systematic literature review on the efficacy of blood transfusions, they would probably have come to the completely wrong conclusion that blood transfusion was an incredibly dangerous procedure. The early researchers in this field hadn't considered the possibility that there might be different blood groups within the population; there was no reason to expect anything so apparently improbable. Nowadays, we take that for granted, just as we take for granted that a key variable in blood transfusion is getting the right match between blood types; once you factor that in, the procedure starts looking like a whole different proposition.

To identify the relevant variables and assess how effective they really are, you need to ask experts both inside and outside your field instead of just relying on your own opinions. One good starting place is to let those experts talk; they're usually happy to explain the reality of their world at length.

For instance, many *basketball players* believe in "hot hands"— that a player can suddenly have a winning streak, when they're able to pull off shots that would normally be beyond their ability. To a *statistician,* however, hot hands sound a lot like someone having the sort of streak of good results that you would expect from ordinary random chance. To a *sports psychologist,* hot hands sound like a plausible physiological phenomenon, even if they don't know what the underlying neurophysiology is. To an *empirical researcher* who listens to the players, it sounds like a hard but fascinating problem to untangle, because players will tell you that if someone on the team is having hot hands, then you pass the ball to them whenever possible, so that they can make the best use of it. That means that even if hot hands are nothing more than a statistical run of chance, the player will be getting better opportunities than usual, so the effect could turn into a self-fulfilling prophecy.

Experts can share some invaluable insights, but that doesn't mean those insights are always right. The geologist telling me for half an hour how to identify a rock sample was honest, but that didn't mean his account was the full story. As psychologists are well aware, the factors really affecting our choices and decisions often aren't the ones we think. The experts also face a lot of practical problems. You're only an expert within your area of knowledge. As we've seen, being an expert diagnostician doesn't automatically make you an expert statistician or surgeon as well. Even within your own area of knowledge, if you're in a field where research is moving with even modest speed, it's effectively impossible to stay on top of the relevant knowledge. That's a big problem. For a lot of fields, even if you were a speed-reader and did nothing each day except read articles, you wouldn't be able to keep up with the torrent.

If you're thinking of reading the literature in *another* area that might be relevant to you, then the prospects get much worse, because you'll probably have a steep learning curve before you can understand the specialist language of that other field. The literature on error is huge, and it comes from a variety of very different fields. Trying to make sense of it all looks like an enormous challenge. Sometimes, though, you can transform a problem into something more manageable by representing it in a different way.

A lot of the human error pitfalls we've discussed are complicated by language. Nonverbal knowledge—what you're carrying around in your head that you don't know you know—is another serious complicating factor. However, there's a way of tackling those two problems in one stroke and helping experts clear logjams in their research. When you turn nonverbal knowledge around, you can produce something nonverbal that is, paradoxically, more rigorous than language: diagrams. Diagrams were to play an important part in the Verifier approach, and in the subsequent Verifier case studies. The next chapter looks at diagrams, and at how the way you represent a problem can make it much easier to handle.

From Words
to Images

EVERY TIME YOU GET CASH FROM AN ATM, YOU are using a machine that has been carefully programmed by its designers to function well in spite of mistakes. Those hole-in-the-wall machines are classic examples of how an apparently straightforward task can go wrong, and a classic test case for ways of tackling those problems.

Think about it: there's only one way of putting in your card correctly, but there are *three* ways of doing it wrong. Every day, millions of people put their cards in the wrong way. So ATM designers think carefully about what to do with each of those wrong options. Do they want to return the card to the user for one option and have the machine retain the card for the others? Or do they want to return the card in all cases, or retain it in all? Above all, they need to make sure that they have covered all possibilities.

Tackling this problem by listing the options in words soon

becomes unwieldy, and it's horribly easy to make mistakes. That approach would look something like this.

Right way up, right way around: ask for PIN

Right way up, wrong way around: return to customer

Wrong way up, right way around: retain

Wrong way up, wrong way around: retain

Even with just four permutations, it's hard to read such a list and check whether it contains any errors or omissions. Here's what that same set of conditions looks like as a table (figure 3).

Changing the representation from words to a diagram makes it a lot easier to check whether all options have been considered and whether each has a decision associated with it. If there's a blank cell in the table when you're expecting all the cells to contain something, you'll spot that in a second. You will spot it even if the table is in a language that you don't speak, courtesy of that nimble, intuitive part of your brain that does efficient pattern matching.

Each discipline tends to use its own set of diagrams. A lot of ingenuity has gone into inventing new types of diagrams and

	Right way up	Wrong way up
Right way around	Ask for PIN	Retain
Wrong way around	Return to customer	Retain

FIGURE 3

better tools for sharing information. Some of the best ways of presenting nonverbal information go back centuries.

This got me wondering whether we could analyze expert errors to figure out which tools were most effective in stopping those errors before they got started. When should you use a verbal tool—asking a better question—the way Gigerenzer and his team did with the Linda problem? When should you use nonverbal tools, like the one in our hypothetical ATM problem?

In this chapter, we're going to look at six tools that let us tackle difficult problems. They're all nonverbal. They all come from academic research, and they all have significant implications for the real world. They also provide a good crash course in pattern matching. The first two, for example, are perfect for helping companies understand the most studied and in some ways least understood research subject on the planet: the consumer. The last of these tools does something for a few thousand dollars that can't be done using the usual approach, even on a budget of millions.

What Card Sorts Tell Us About How Women Dress

My colleague Sue Gerrard investigated perceptions of women's work dress as part of her master's thesis. Specifically, she asked, "What signals does a woman think she's sending out when she wears a particular outfit, and what signals are people perceiving from that same outfit?" The topic had been investigated by previous researchers, who typically found that sex and power were the two most common variables transmitted. Sue wondered whether she might find new insights if she used a different method from the ones used by those earlier researchers.

Most of the previous research on how women dress at work had used interviews or questionnaires. Sue wondered what she would find if she used the card sorts technique that I had used at Nottingham. Each of the cards in her pack had a different photo of a woman's outfit on it. She asked the people in her study to sort those cards into groups of their own choosing. Each person sorted the pack as often as they wanted, using a different criterion each time, and using as many or as few groups as they wanted each time. What Sue found turned into a classic paper.

Her first main finding was a new variation on an old theme, namely sex. About half the men in her study, but none of the women, *voluntarily* chose to group the pictures according to whether the women in the pictures were married or unmarried. Sue had edited the photos so that they only showed the women's clothing; hence participants would not be influenced by the models' faces. There was just one exception: one model had her left arm posed across her body, so it couldn't be easily edited out; a wedding band was visible on one finger. You might think that this would be the only card that the male study participants added to the "married" pile, but you'd be wrong. These males were cheerfully putting cards on the "married" pile even when they couldn't see the models' faces or arms, let alone a ring finger. Clearly they were assessing each woman's marital status by the way she was dressed. Sue's study was limited to the card sorts investigation, but if she'd had the time, she'd have used another technique to drill down to the reasons for this categorization. You might guess that if a male thought the outfit was provocative in some way, he would infer that the woman was single. An unprovocative outfit may have been interpreted by those men to mean that the woman was married. Sue's findings had a lot of implications for clothing designers.

But there was also a deeper, subtler finding. In each round of the study, people were allowed to sort the cards into however many groups they wanted. The women would sometimes sort into two groups, or three, or four, or sometimes more. But the men tended to sort the cards into just two groups.

Sue's results had far-reaching implications. Looking at the results, you had to wonder if the males were sorting the cards into two groups because they were using simplistic, binary, yes/no thinking about the world in general, rather than because of something specific to women's working dress. If it were a case of simplistic thinking, then it was worrying, because most of the decision makers at the head of governments and major companies are male, and most questions in this world don't have simple yes/no answers. She may have unexpectedly tapped into that big, scary picture of decision makers making life-and-death judgments based on simplistic, binary reasoning. That insight made her thesis a key paper among researchers using this variety of card sorts.

It was several years before anyone was able to answer one of the questions raised by her paper. Over those years, researchers using this version of card sorts patiently built up a library of data that they shared with other researchers, in areas as diverse as web-page design and risk assessment. At last, a researcher at the Open University in the United Kingdom, Linda Price, pooled those results and concluded that the finding wasn't simply connected to gender. Also, it wasn't simply a case of women being experts in women's clothing, and therefore using more categories.

The full picture was much more complicated. The way people sorted the cards had something to do with our old friend, expertise. In some cases, experts tend to sort things into more groups than novices do; in others, they sort them into fewer categories because they're focusing on a small number of key issues, whereas novices are dragging in a large number of superficial issues.

It's unusual for a master's thesis like Sue's to achieve so much. Part of its success was because she was asking an insightful question. The other part was because the tool she used made it easy to *answer* the question. Asking questions is easy; asking good questions is difficult; asking good questions in a way that makes it easy to answer them is a rare skill.

The next case study also involves a different way of asking questions, and of representing the answers. It involves a topic that we will explore in more detail later in the book, because of the way it can skew the judgment of otherwise rational, hard-headed individuals. The topic is perceptions of attractiveness.

The Signals Your Website Is Sending

Anyone who makes or buys software is obsessed with *design,* but there are two different meanings of the term. Sometimes it means the superficial appearance—the software equivalent of slapping a new label, paint job, or racing stripes on an existing product and pronouncing it "New and Improved!" But more significantly, software design also refers to decisions about what the product does and how to make that product easy to use.

Google's *superficial appearance* couldn't be simpler: it looks like a blank screen with a box in the middle of it. But when you type in your keywords, the *functional design* helps you find lots of records very fast. Ultimately, the most important part of

Google's design is not about the bland box or the colors on the home page. It's about a major company figuring out what their users want and putting a lot of effort and brains into making their product give those users what they want as smoothly and efficiently as possible.

Measuring the effectiveness of a commercial or industrial design is therefore a big deal. If you're a manufacturer, you want to have trustworthy numbers from your product and market research teams before the product is launched. One key goal is that you want it to be as attractive as possible, with as few un-attractive aspects as possible, as well as being functional. That looks mind-numbingly obvious. It implies that you're dealing with two concepts that are opposites of each other, *attractive* and *unattractive,* which also seems incredibly obvious. But they're not. As our next example shows, the human mind can perceive something as being both attractive *and* unattractive at the same time. One of my students, named Zoe Szymanski, discovered this as she investigated a question as elegant and counterintui-tive as Zeno's paradox: *What if the opposite of "attractive" isn't "unattractive"?*

Think back to the last time you were stopped in the street by a market researcher with a clipboard and a list of questions. Probably you were asked to give an answer based on a scale. The question might have been something like, "On a five-point scale, where 1 means 'unattractive' and 5 means 'attractive,' how would you rate our product?" This type of survey question is known as Likert Scales. They're named after Rensis Likert, an American psychologist who invented them in the 1930s, and who reckoned that he was onto a very big thing. He was right. Eighty years later, they're still going strong. *Everyone* uses them. If you're launching a new product or a new website, and you want to be sure that it's getting high ratings for attractiveness, usability, effectiveness, cost or a hundred other variables, you'll probably be using Likert Scales on the prototypes before the real thing goes live.

There's a huge body of scientific literature on Likert Scales, and another on something called measurement theory; people in those fields have spent lots of time investigating how you should word each end of the scale and whether the scale should run from a negative number to a positive, or from zero to a positive.

As often happens, there were some questions where the answers looked so obvious that nobody paid them much attention. Zoe's was one of those. The assumption that something is either attractive or unattractive is not only wrong—it's worse than wrong: it's *nearly* right, but not quite.

Zoe was aware of that issue because she knew about previous research that had investigated a very similar underlying question. Most people tend to think of masculinity and femininity as being on opposite ends of the same scale. Androgyny theorists wondered what happened if you thought of them instead as *two separate scales,* each running from low to high. Even with simple stereotypical concepts of "masculine" and "feminine," you could see patterns that were obscure before. The archetypal plaid-shirted lumberjack and the busty blonde country-western singer each have a clear place on the scales. The lumberjack is high on the masculine scale and low on the feminine. The Dolly Parton look-alike is high on the feminine scale and low on the masculine.

Using two scales meant there was also a place for people who were *low* on stereotypical masculinity and femininity, such as the sort of people who are mocked for being dull, nerdy geeks. And we also find slots on the scales for people *high* on both stereotypical masculinity and femininity. Elizabethan noblemen, who were gorgeously dressed and wrote beautiful poetry, also happened to be murderously violent and some of the most rapacious plunderers in history.

Zoe asked people to look at various website home pages and then answer a set of questions, which included one crucial pair of questions. Specifically, she asked how much each home page *encouraged* people to go on into the site, and how much it *discouraged* people. This a major question in website design, because visitors to a site typically decide in less than two seconds whether they will stay. If you can't entice them in within two seconds, the rest of your website might as well never have been created.

She found that in most cases, the "encouragement" answer really was the opposite of the "discouragement" answer. When people found the site highly encouraging, they ranked it low in discouragement. In quite a few cases, though, the "encourage" and "discourage" ratings were both low: the page was simply boring, not interesting enough to tempt people in, but without anything actively and conspicuously wrong.

But in a few cases, the answer was high on *both* scales—high on encouragement, high on discouragement. At first glance, that sounds downright bizarre. It means that a visitor could feel simultaneously attracted to a website *and* strongly put off by it. Once you stop to think about that, though, you can see how it could happen. For example, a holiday destination that describes itself as "exclusive" sends out an encouraging signal of luxury but a discouraging signal of expense, and possibly snobbery. A page could be sending out a strong mixed signal like that, but you would not have picked it up in the market research unless you asked the question correctly (figure 4).

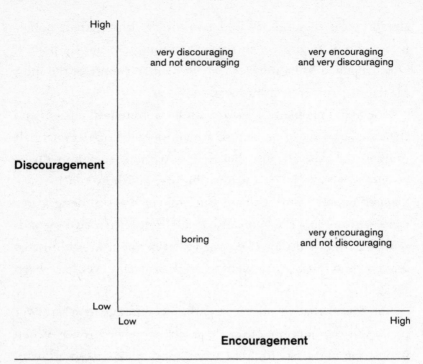

FIGURE 4

No matter how you look at it, if you want to get good information out of people and you want to solve difficult problems, it helps to do two things really well: ask a question no one has ever asked before and figure out the right way to ask it—which may involve not using words.

Sue Gerrard did this with her study about women's clothing. She supplied the pictures; the participants supplied the words and concepts. Zoe Szymanski did the same when she used the approach of *paired scales*. Scales use some language, but they are also strongly visual. Respondents don't have to struggle to articulate how they feel. They intuitively understood how scales work, which results in cleaner conclusions that are a better match with reality.

The Complexity of Axes

One of the biggest drivers of national prosperity is innovation. Think of the effect the Internet has had on everyday life, or how some of the wealthiest organizations on the planet, like Google and Apple, are based on technologies that were invented within our lifetimes. At first glance, it looks as if those innovations involve massively more complex technologies than their predecessors, but when you think closely about the issue, you realize that the reality is more obscure. Governments want to foster innovation, but they have a lot of trouble working out just what innovation actually is, and whether complexity in an innovation is a good thing or a bad thing. If you can't define something, it's difficult to manage it, and that's a problem when you're talking about concepts as fundamental to economic strength as innovation and complexity.

Take watches, for example. A mechanical watch has a huge number of moving parts, whereas a digital watch has no moving parts. You could argue that the digital watch is less complex than the mechanical one because it contains fewer separate parts, or you could argue that it's more complex because of the sophisticated structure of those few parts.

When you can't even be sure whether a technology as ubiquitous as a watch is becoming more complex or less complex, looking for regularities in innovation is a serious challenge. However, it's a critically important challenge if you're trying to foster innovation, which is a key issue for governments and individual organizations. They spend enormous amounts of money trying to foster new technologies, but many of the underlying principles are still obscure. I was interested in one particular question related to this issue, namely how to measure complexity. I was interested in it for two reasons. One was a desire to tidy up a messy problem. If we had a clean, systematic way of measuring complexity, researchers could reexamine the technological record to see if there were any underlying regularities in how technologies changed over time in regard to complexity. The other involved identifying strategic dependencies. In the increasingly interconnected global economy, an apparently trivial glitch in the supply of a single raw material or component can have dramatic repercussions around the world. The approach I was planning to try for measuring complexity would, I hoped, throw some light on that problem as well.

Some of the first people to try to measure complexity were anthropologists attempting to describe the growth of technology and civilization. They ended up getting their fingers burnt. Victorian colonial researchers who visited far-off lands tended to take a patronizing view of "the natives" and their tools as being primitive. That view was gradually abandoned, as art historians started to grasp the underlying sophistication of the conventions behind these art forms, and researchers of technology realized that often the best solution to the problems faced by those cultures was actually a simple solution. The concepts of "primitive" and "advanced" fell largely out of use, but it was still clear that some products were more complex than others, even if nobody was quite sure how to measure the difference. It was a classic case of a gut feeling in search of a solid method.

In 1976, Wendell Oswalt, an anthropologist at the University of California–Los Angeles, proposed using the concept of the "techno-unit" to measure the complexity of an artifact. By his definition, a spear made from a piece of wood sharpened at one end is a single techno-unit; a spear with a stone head attached to a wooden shaft using some rawhide consists of three techno-units (the stone head, the wooden shaft, and the rawhide).

The techno-unit was a good start, but it had obvious limitations. For instance, by this measure, a spearhead made from a broken piece of stone picked up on the beach would have the same complexity measure as a Solutrean spearhead that was made using skills only a few humans have mastered. Your "beach" weapon would also have the same complexity as a spearhead fashioned from the finest quality steel by a master Japanese weapons maker. Each of the three spearheads would count as just one techno-unit.

When I started looking at the problem, I realized that a possible solution had been sitting unused for over two centuries. It's an approach from mathematics, and it's called graph theory. It was invented by a Swiss mathematical genius called Leonhard Euler. Today graph theory is a major area of mathematics that is at the core of modern inventions like the Internet.

Inventing graph theory took Euler a whole afternoon. The story of how he invented it is well known in math circles. In the early 1730s, the locals of a town called Königsberg were in the habit of taking afternoon strolls beside the river and across the seven bridges joining the two riverbanks and two islands in the middle of the river. They began to wonder if there was a route that crossed *all* the bridges just once, without missing any or crossing any twice. Try as they might, they could not find a solution. The problem seemed unsolvable.

Euler set about solving it by using a favorite technique of experienced researchers. Instead of taking on the small, manageable-looking question about the bridges, he took on a much bigger problem about routes in general. If he cracked that, he would crack the bridges problem, but he would get a huge set of extras in the process. That may sound like a bizarre piece of reasoning, but in fact it's often easier to solve the bigger problem than the smaller one. Euler's story is a classic example.

First, he stripped the problem down to essentials. The *real* problem, the deep structure, wasn't about bridges and islands and riverbanks: it was about routes between points. So he drew diagrams, reducing the islands to points and bridges to simple lines, and analyzed the problem that way. That's what gave his solution so much power: it was about underlying principles, which could then be applied to a wide range of problems. These days, experts use graph theory to solve problems such as how to route Internet traffic, air traffic, and road traffic, as well as how to schedule complex construction projects and model social networks, and a ton of other applications.

Graph theory is composed of elegantly minimal parts. There are just two, and they're variously known as arcs and nodes or edges and vertices. Whatever you call them, they can be represented as a set of points with lines between them. Figure 5 provides an example. The three circles are the nodes, and the two lines are the arcs.

In this graph, the top node is connected to both the other nodes, but the two other nodes aren't directly connected to each other. It's a deceptively simple-looking method. In fact, it's extremely powerful. With his solution to this bigger problem, Euler was easily able to demonstrate that the problem of the bridges was unsolvable; there really was no way to cross the seven bridges without either crossing one bridge twice or missing one of the bridges. The demonstration takes just a few minutes.

But that was just the beginning. You can do a huge number of other things with graph theory. For instance, you can identify

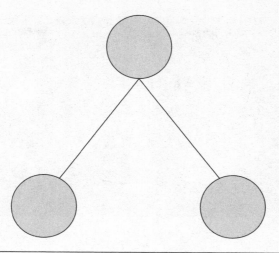

FIGURE 5

the shortest route between any two nodes; you can also identify the next-shortest route if one of the nodes on the shortest route is out of action. That's invaluable for routing Internet
traffic, where you're trying to distribute the load as efficiently
as possible.

You can also use this approach to model how complicated
it is to build a particular product. Instead of showing links between islands and riverbanks, you're showing the links between
the product and the tools and materials needed to produce it.
That's based on a concept known as the object's "fabricatory
depth," which spells out all the items needed to make a given
object. Here's an example. It involves one of the earliest human
inventions, a Stone Age tool known as the hand ax (figure 6).
To make the oldest style of hand ax, you need a piece of flint (as
raw material, figure 7) and a rock, called a hammerstone (as a
tool to bash the flint into shape, figure 8). For a more *advanced*
hand ax, you must go down to a deeper layer. You still need a
piece of flint, and you still need a hammerstone. But you also

FIGURE 6

FIGURE 7

FIGURE 8

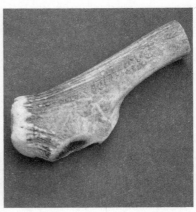

FIGURE 9

use a soft hammer to do fine work. The usual soft hammer is a chunk of deer antler (figure 9), which you work to size and shape with a hammerstone.

Here's what happens when you show the production process for these two types of hand ax with graphs (figures 10 and 11).

The pictures tell us at a glance that the new-style hand ax is slightly more complex than the old-style ax. With the old-style hand ax, you need two boxes to show the underlying items;

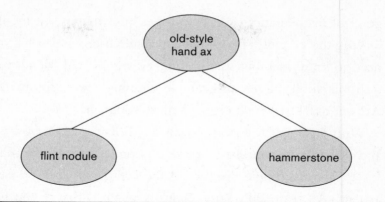

FIGURE 10

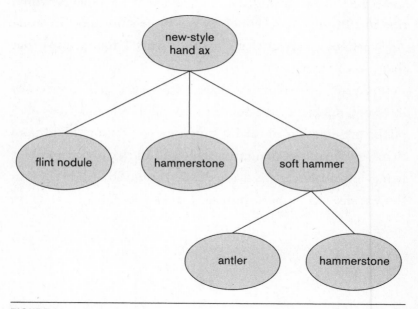

FIGURE 11

with the new-style ax, you need five boxes. You can actually count the levels or the number of boxes to quantify an object's complexity.

If you want to get sophisticated, you can then extract all kinds of measurements from the graphs. But even people whose grasp of mathematics is shaky can get information just by observing the graphs. Those graphs will often tell you what you need to know without using numbers, via our old friend, pattern matching. Here's an example without words (figure 12). Answer quickly: which object is more complex?

With this format, it's easy to see the differing layers of depth and the differing numbers of arcs and nodes in each diagram. It's a rigorous, simple measure of the complexity behind each artifact. All reasonably sane scientists assumed that if you had some way of measuring the complexity of technology, you'd find that Bronze Age technology was more complex than Stone Age technology. And that's roughly what I found, too—but there was a twist.

In April 2007, I first presented this work at a conference in Southampton, in the south of England. Showing data in a public presentation is usually a nightmare—for anything more than very small sets of figures, there's a real risk of the numbers being unreadable at the back of the room. So I didn't try to show tables of numbers. Instead, I stuck paper diagrams up on

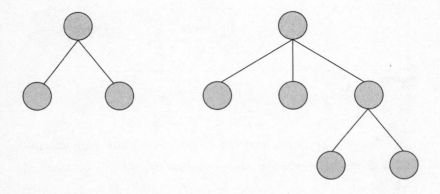

FIGURE 12

the whiteboard, showing the graphs for different types of hand axes and axes through prehistory. For the Bronze Age data, it was a big diagram. For the hand axes, in contrast, the diagrams were practically invisible from the back of the hall. That was the point. They were ridiculously simple, as you've just seen. The key point was to show how simple they were, and the relative size of the diagrams made that clear without needing to get into the precise numbers. The finding that took me by surprise, and also surprised the audience, was where the graphs showed a sudden rise in complexity. It wasn't in the Bronze Age, which is what everyone had expected. It was much earlier, in the Neolithic, when people started digging deep mines to get at the highest-quality flint. Those mines brought a huge increase in the technology needed to make an ax. Yes, the Bronze Age axes were even more complex, but a lot of the groundwork for that complexity had been done much earlier, with those Neolithic mines.

Besides me, two other researchers at the conference had also discovered how formal representations could solve the complexity issue. Both of them had tackled problems that I had steered clear of, and both had done brilliant jobs, each using a different formal representation from the graph theory I had used. None of us was aware of the others' work, but the three talks couldn't have been a better fit if they had been planned that way in advance. And it had all come out of one of the simplest notations imaginable.

Our next example looks at a problem caused by having vast amounts of information. It involves applying the concept of pattern matching to the problem of finding relevant information when you're doing an online search.

Visualizing Online Searches

A key challenge with online searches is constructing the algorithms—the instructions that tell the computer how to find and prioritize relevant records from among the enormous number of records that are out there. If you search for the keywords *John Smith Pocahontas,* for example, the search engine will take less than a second to identify 2.3 million records that might be relevant.

The people who design online search software know a lot about the problems involved in online search. They could tell you that the average number of keywords a user types in during a search is 2.4. The designers could also tell you the average number of typos in a search, which they have to figure out a way to circumvent. But these are not the *only* problems, and there's one big problem that most people know all too well, namely working out which records really are relevant. The search engine companies have algorithms that try to do that, but they're not brilliant, and there's a good chance that the record you want is buried somewhere far beyond the first page of results. Once the search engine finds those 2.3 million records, it in effect tosses those answers into your lap and says, "Your problem now."

Look at what just happened. You went from harnessing the power of a search engine so powerful it can identify 2.3 million potentially relevant records in under a second, to reading at a speed of maybe a couple hundred words a minute. You can bump up the speed a little by skimming results or by using the "find in file" function to jump to the next keyword in the file. That might take you to the giddy heights of five or six hundred words a minute, about as many words as are found on a page in an average paperback. Even if you only have a modest number of hits—say, a hundred thousand—you're still confronted by probably a million pages of text. If you read them at a rate equivalent to five paperbacks per day, it would take you about three years to read through all of them. In principle, the most relevant results should be at the top of your list. The reality is a lot more prosaic. Yes, that often happens. But the ranking can easily be skewed.

Often the search software can't tell what would make one record more relevant to you than another: it just lumps together at the top all the searches that contain all of your keywords, chunked by characteristics such as how recent they are. If you're searching for someone with a name like Jim Smith in Detroit, you're stuck wading through a huge number of potentially relevant records. Heavy-duty researchers sometimes deal with this sort of problem by using the guideline, Take the obvious and reverse it. If the obvious approach is to find relevant records by reading what's in them, one reversed solution would be to find relevant records *without* reading anything. That may be unusual, but it's far from impossible. That's where diagrams and parallel processing come in.

Imagine that someone has gone through one of the records turned up by your search, highlighting your search terms wherever they occur. If you look at it from a distance, it might look like figure 13.

Does that tell you anything useful? Yes, in fact it does. The swath of light gray squares represents all the words on the page. The single dark gray bar tells you that there's only one mention of your keyword in that record, and the mention is very early on. That's a start. Figure 14 provides another example, showing another page with the occurrences of your keyword

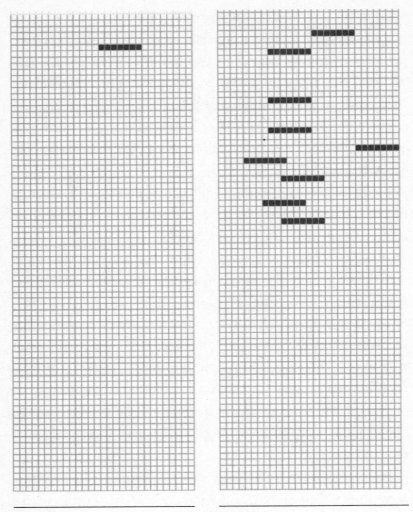

FIGURE 13   **FIGURE 14**

highlighted throughout. Which of these pages contains more mentions of your keyword?

The answer is obvious: one record contains a lot of mentions, whereas the other only contains one. That only takes about a second to work out, without even having to open the records.

So, without reading a single word, we can already tell quite a lot about relevance. But there's more.

Look at figure 15. Comparing these two web pages, we can see that one record contains mentions of the keyword all the way through, whereas the other contains mentions only in two sec-

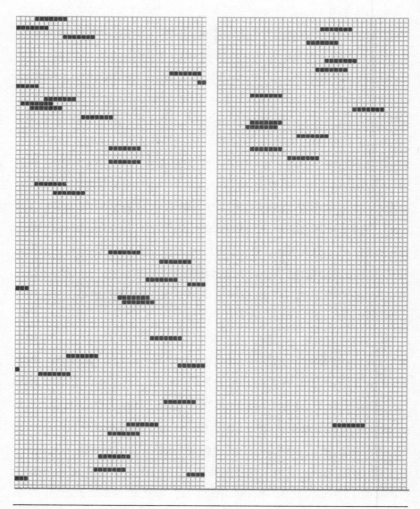

FIGURE 15

tions—a cluster at the beginning and a single mention at the end. Without having to scroll through the entire document, we're seeing structural information. That's hard to make sense of if you're reading a document of significant size, and it's something that search algorithms based on "proximity" don't touch. They only search for records that contain two specified words within a specified distance of each other. They can't tell you about the structure of the article the way the above images do. And there's more. If you put all this together and include color coding (or shade coding for the color-blind), you can do some pretty sophisticated interpretations of a batch of links that you've found with your search engine. Figure 16 shows what we got from a search about medication for bradycardia (unusually slow heart rate).

This shows the content of the first ten links we found in a search using one of the major search engines. The *medication* keyword is in light gray, and the *bradycardia* keyword is in dark gray. Which of these records are highly relevant, moderately relevant, or irrelevant? It's obvious at a glance that record 1 is the most promising. It's also obvious that some of the web pages are long—we'd need to scroll down to see the entire page—while others are comparatively short.

So you've just assessed the relevance of ten records totaling several thousand words in a matter of seconds, without reading a single word. That's starting to look seriously useful. But there's still more. In that last example, the search was actually for *medikament* and *bradykardie*—you've just assessed the relevance of ten records in German. If you're a working scientist trying to decide which foreign-language documents are worth running through a translation program, that's a big deal. That's even more useful for national security agencies, who often urgently need to process huge quantities of captured documents but are

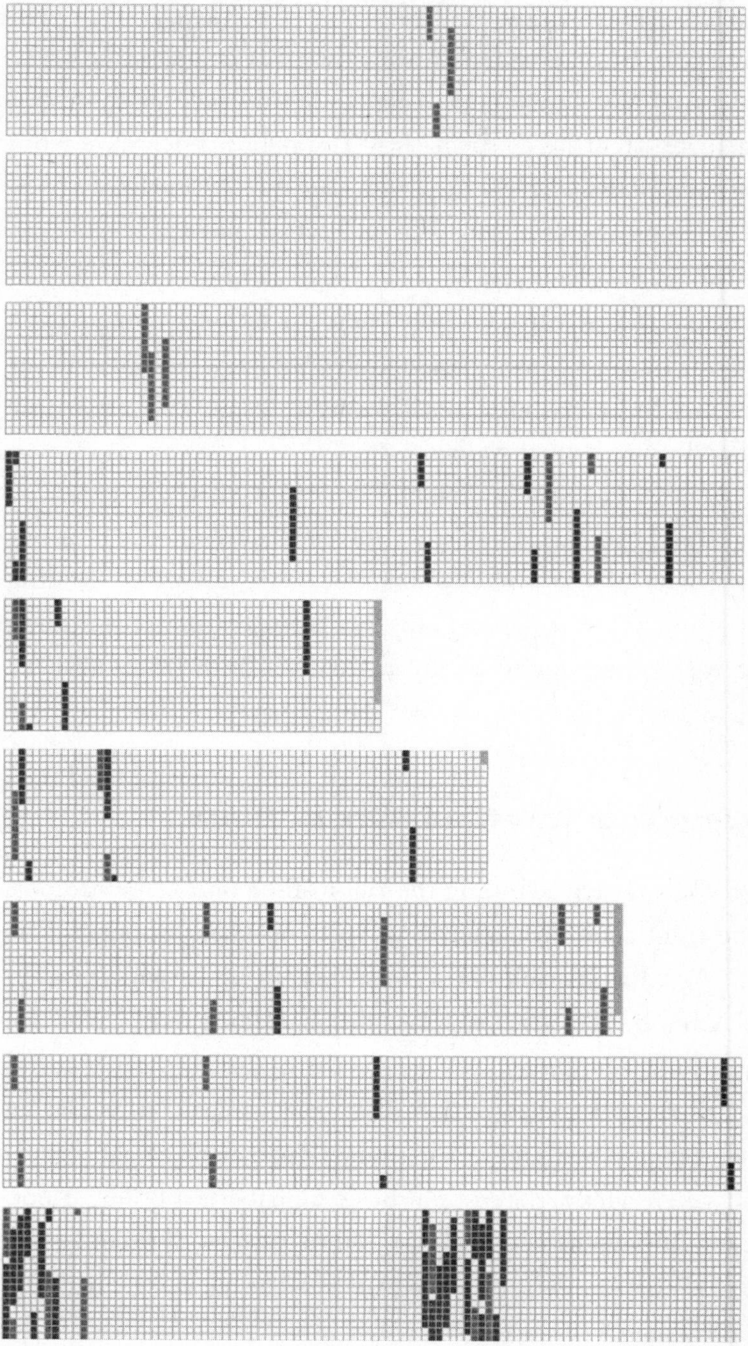

FIGURE 16

limited by the scarcity of staff who speak the relevant languages. Visualizing the records in this way, they can greatly reduce that problem by having ordinary agents identify records that need immediate attention by specialists. The police officers we work with particularly like the way you can use this technology to search through large numbers of witness statements to spot possible links between cases.

It's a classic example of how asking a nonobvious question can give you a very powerful, nonobvious answer. Obvious answers are cheap; nonobvious answers that are also useful are rare, and usually much valued.

The next case study involves solutions that are practical, cheap, and powerful. They were also obvious, but only in hindsight. Many of our previous examples have involved "finding your way" as a metaphor. This one involves literally finding your way, when you're trying to reach a destination and the signage just isn't helping you get there.

Finding Your Way from Here to There

Have you ever tried finding the right office on a large campus, or the right room in a sprawling building like a hospital? The classic bureaucratic solution is to add more signs, which results in a classic bottleneck, the "T-junction reading wall." You see these in most hospitals. You get to the end of the main corridor leading past the reception desk, and you're confronted by a wall covered with signs for every department in the hospital. If you're lucky, they might be in alphabetical order; if not, you might have to read through ten or twenty of them before you find the one you want, while busy hospital staff back up behind you.

There's also a fair chance that the name on the sign is out of date. A famous example is the oldest bridge in Paris, built in 1607; it's known as the Pont Neuf, which literally means the "brand-new bridge."

So, if adding more signs just overwhelms people with more information to process, what can you do to help people avoid getting lost? The answer involves a toolbox approach—there are several methods, which all complement each other neatly, so you can mix and match to suit the problem.

Here's an example I was involved with. A university wanted to get visitors where they needed to go on campus. The campus was huge. They were spending a fortune on signs and wanted to come up with a simpler, systematic solution. Our team wondered: could we figure out a way to get people around campus *without* signs?

We had asked the better question. When I began looking into the problem, I found that there were several mini-roundabouts on the approach route from the nearest main road. These led to a key junction, where most visitors will either go left to the main reception building and main parking lot, or right toward the sports center. Unfortunately, this important junction was situated at another mini-roundabout and wasn't obvious to a stressed visitor arriving late in the rain. It was easy to go straight through and end up either on a dead-end road or in a parking lot, with no easy route back to the twenty-four-hour reception. There were lots of signs on the approach to this last mini-roundabout, far too many to read comfortably without slowing the car to a crawl. When you did eventually reach your destination, the footpaths were a tangled network without many distinctive landmarks. Most of the buildings looked depressingly similar, so getting to the right building would be a challenge.

In short, it was a mess, a great maze for any rat-runners in the psychology department, but not exactly great for lost humans. I had already done an analysis of the network of roads and walkways for another project. Cribbing from Euler's bridge problem, I had found that the road only had three main decision points: one at the last mini-roundabout, one at the junction that leads to the sports center, and a final decision point at the far end of the sports center road. That meant you could install just three landmarks and they'd be enough to get people to the key destinations. I also found that 80 percent of the buildings on the main campus could be seen from the main footpath, which is easily visible from the last mini-roundabout, the main reception, and the main parking lot. That told me which places needed most attention; the question was what sort of solution was needed for each place.

The Cheaper, Visual Solution

For a start, you need a nonverbal way of telling travelers they are about to approach their most important mini-roundabout, the one where they will need to start making decisions. A simple approach is to erect a distinctive piece of sculpture. You can't put it in the middle of a mini-roundabout, which is just a small paint circle in the middle of the road, but you *can* put it on the grass at the far side, where it will be easily visible. You can then simply tell visitors that when they reach the roundabout with the statue, that's where they have to turn.

But the problem is not completely solved yet, because a surprisingly high proportion of people have trouble telling left from right. People also tend to forget directions after a few seconds. We've all been there. By the end of a long drive, the human memory can be forgiven for getting confused, even if a person *can* tell the difference between left and right. There's a simple solution, though. *You build the correct direction into the statue.* In this case, I suggested commissioning a statue of birds in a tree, with all the birds looking toward the main reception and parking lot. Then the visitor just needs to remember one rule: "When I reach the birds statue, I let the birds show me where to go." The instruction is unusual, distinctive, and memorable, and it doesn't depend on knowing left from right.

But what do you do for visiting sports teams who need to reach the sports hall, which is in the opposite direction? Answer: You tell them that when they reach the birds statue, they should look for the junction with the statue of a giant football. That's easy for coaches and players to remember, and there's no way it will be confused with the statue of the tree and birds. You can put another statue, of a discus, at the other end of the sports hall road, to mark the other decision point.

The signage is all low-cost so far and is easy to remember—or to guess, if you're arriving without previous instruction. But it's not the whole story. Anyone who has worked with academics will know that sports teams sometimes stir up strong feelings in other departments, usually feelings of contempt. Statues of a football and a discus may not go down well with faculty and staff; the statues imply that sports are more important than academic departments. There's an elegant solution: have the sculptor make one statue based on a diatom—a microscopic animal whose intricate shell looks a lot like a football. That statue would look enough like a football to help a visiting sports team find its way, but it has a whole layer of educational subtext that will keep the biologists happy. For the other end of the road, the sculptor can make a statue of the Phaistos disk—an ancient

Greek artifact, covered in undeciphered writing, which looks enough like a discus to help a lost sports team but has enough backstory to keep the classicists, linguists, and cryptographers happy. When the statue is unveiled, you can even challenge visitors to try cracking the mystery of those ancient symbols.

This brings us to the problem of walkways leading to the various academic buildings. The obvious solution was to do something distinctive to the main path, which was visible from most of the central buildings of the campus. When I mentioned this to the person in charge of the grounds, he looked like a man who had just been given an unexpected birthday gift, and told me that he had been planning to resurface that path within the next few months. Resurfacing it in a distinctive lighter color would be easy and wouldn't cost him anything extra. For most lost-visitor queries, you could simply direct them to the light-colored path, and give them directions from that: "Follow it till the end" or "Follow it until you see a building with a brick circle just under the roofline."

Taken together, this handful of changes would reduce or eliminate about 80 percent of the lost-visitor problems on campus. Total cost? About ten to twenty thousand dollars.

Compare that to the total cost of installing a complete set of traditional verbal signposts on a big hospital or university campus: about a million dollars. No wonder the client was happy to hear the figures for the nonverbal signage design. The nonverbal signs wouldn't replace the verbal signage—that was out of the question—but they would reduce the number of lost visitors, giving them a better experience of the campus. It might also mean that the campus planner could reduce the number of verbal signs (and maybe make an overall saving on his original budget).

There was just one potential caveat: when I mentioned the light-colored path to people, they all suggested the same color. If the university does adopt that design, helpful university staff will ever after be telling visitors, "Follow the yellow brick road. . . ."

So, in the end, the visual representation of knowledge makes a difference. These six handy tools help us solve problems because they exploit our human gift for pattern matching. Now let's see what happens when we add in a century-old mystery and a few all-too-human errors to the mix—mostly mine.

Ambush In Your Mind

KEEP A REPLICA HAND AX ON MY DESK AS A PAPERWEIGHT. It's beautifully colored and textured by nature, its flint stained into rich reds and ochers and yellows by iron in the soil. It's oval in shape. One end is unworked, just the rounded end of the naturally rounded beach stone that it was made from; the other end has been worked into a pointed *V* shape.

There are millions and millions of real hand axes in the world; our early ancestors started making them long before modern humans evolved, and kept on making them according to the same basic pattern for hundreds of thousands of years. I made this replica myself. It's a reminder of the chain of expertise that was passed from human to human, unbroken through time, continuing even across changes in species.

The ax is irresistible to visitors. They want to pick it up and hold it. A lot of them respond the same way, with variations on "You can see how it was made for . . . ," followed by completely different ideas about how it was obviously meant to be used

right-handed, or left-handed, or ambidextrously, or for stab-
bing, or for cutting. I stopped trying to argue with them years
ago. Now I just smile politely and move on.

The question of how or why humans became left- or right-
handed is one of humankind's oldest mysteries. With some
strange exceptions in the animal kingdom, we're the only species that displays a preference for one hand or the other.

One of the big questions about human handedness is when
it began. Scientists wonder if only *Homo sapiens* developed this
preference, or if it occurred farther back in our family tree,
among extinct species such as *Homo erectus* or australopithecines
or Neanderthals. If we knew the answer to that question, we
would immediately get insights into other big questions, such as
whether handedness is linked in some way to our other distinguishing characteristics, such as tool use and language.

Unfortunately, clues for early human handedness are hard to
find. Many objects—baskets, wooden tools, clay works—that
those hominids may have made have long disappeared, and the
skeletons of early hominids are so rare and incomplete that we
can't tell from looking at them if the population at a given time
was predominantly right-handed, like modern humans.

If archaeologists were to have any hope of cracking this
puzzle, the answer would have to come from objects that have
survived in large numbers. That makes stone tools—the originals of the one sitting on my desk—the best bet. But after looking at a lot of examples, archaeologists researching handedness
eventually came to the same conclusion: the shape of a stone
tool doesn't really help determine whether it was intended for a
right-hander or a left-hander.

When I worked in field archaeology, I got lucky and spotted
a possible key to the puzzle. My discovery came at a time when

I had no real experience of serious research, and as a result, I made a mistake. I could easily have avoided it, but I didn't, so it was several years before the answer came out. At one level, that wasn't a huge deal. It wasn't a medical discovery that was going to save lives. The delay amounted to a few years, not decades. At another level, it demonstrates how small things can affect how science unfolds. Exactly the same things could have gone wrong if it had been research into Alzheimer's or cancer. That's one reason for telling this story; it gives an insider's view of the realities of research.

There's also another reason. This piece of research taught me something important. It gave me firsthand experience of finding something that had been staring experts in the face for more than a century but had been overlooked all that time. When you've had an experience like that, you're a lot more ready to take on research challenges that might otherwise seem impossibly daunting—challenges like the Voynich Manuscript.

Stone Tool Making 101

Like a lot of other skills, making a stone tool is built on a handful of simple underlying principles. It's about hitting a brittle rock (usually flint) with a tough rock (the hammerstone) that won't shatter into razor-sharp pieces in your hand. If you hit the flint the right way, you can safely predict how big a piece will flake off, and where it will flake off. That flint flake will always have some key features because of the way flint fractures. Figure 17 provides a rough sketch of a flint flake that has been detached from its parent piece (the core). Flakes like this can be any size between about an inch and about a foot long; they vary from chunky to wafer-thin, but the key features remain the same.

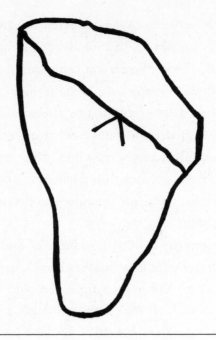

FIGURE 17

We'll use the inverted *V* shape as the point of reference. Its tip marks the spot where the hammerstone struck the flint core, detaching the flake. The top surface is known as the *striking platform*. The inverted V embedded in the flake is caused by the shock wave from that hammerstone blow. It's called the *cone of percussion*. Most flakes have one. It's one of the standard features of a flint flake. It goes deep into the material of the flake, and therefore stands up well to weathering. I've handled flint flakes that were hundreds of thousands of years old, stained right through with yellow and iron minerals over the millennia, whose cones of percussion were as clearly visible as the day they were made, even though the edges of the flakes had been gently smoothed by long abrasion from river sands.

The cone of percussion is usually perpendicular to the striking platform. When someone hits the flint with a hammerstone, the resulting shock wave goes straight into the flint symmetrically. Sometimes, though, the cone of percussion is skewed to one side. The usual explanation is a mis-hit with the hammerstone, but nobody seems to have given the matter much thought beyond that.

I wondered why that mis-hit might have happened. More specifically, I wondered whether the angle of the mis-hit might have something to do with handedness. I didn't get too hung up in detailed speculation about just how that might work—I knew enough basic physics to know that the details were likely to be horribly complicated. It would be simpler just to cut to the chase and test the idea with some flints and flint "knappers"— people who know how to make stone tools.

I worked on the project with my wife-to-be, Alison, who was a professional archaeologist. She supplied the archaeological knowledge; I was just starting a PhD in experimental social psychology, so I scrounged knowledge about how to design experiments and analyze statistics from my PhD supervisor and anyone else who would stand still long enough to answer my questions. We tested a batch of different people, left-handers and right-handers and an ambidextrous person, and we found that there was indeed a link between which hand these people used to make their tools and the types of marks they left in the stones. The correlation wasn't perfect, but it was way above random chance levels. If a cone of percussion was skewed, the skew gave a strong clue about whether that flake had been struck by a right-hander or a left-hander. That result changed the game. It wasn't the end of the story, by any means. Someone would have to replicate our study on a much larger scale; it would need to be checked by measuring the angle of the cone of percussion on flakes that were a few thousand years old, when the proportion of right-handers was much the same as today; then it would need to be applied to really ancient flakes. But it was a great start.

Word on the Street

We were getting ready to write the paper on our findings when a colleague dropped by with some news. Nick Toth had come up with a really neat method for inferring handedness from flint flakes. It used the cortication pattern on the flakes.

The method developed by Toth, then a researcher at Berkeley and now codirector of the Stone Age Institute in Bloomington, Indiana, was utterly different from ours, and from the description we heard, it was far better. For one thing, its criterion was easy to use, unlike the cone of percussion, which can be fiddly and tricky. High-quality flint occurs as glossy black nodules ranging from the size of a softball to that of a basketball; weirdly, that glossy black material is surrounded by a white, chalk cortex, or skin.

When you start removing flakes from a flint nodule, the first flakes to be removed often have some cortex still visible on their backs. Toth had homed in on the stone's color—black and glossy like obsidian versus white and chalky—an easy method for even the freshest novice to apply. The killer, though, was that his method sounded 100 percent accurate; ours was nowhere near that.

Hearing this, I decided that there was no point in trying to publish a method that was nowhere as good as Toth's sounded. Even if our paper was accepted for publication, it would make us look like second-raters by comparison and clog up an already overloaded literature with one more paper that nobody would bother to read. The data and the flakes from our study went into boxes and gathered dust, throughout my marriage and divorce and years afterward.

Deciding to shelve those findings was a mistake—a big one—but it wasn't a stupid one. Academics, more than most, know all about information overload. You simply can't read everything that is relevant to your field; there aren't enough hours in the day. To decide which papers to read, you use heuristics, rules of thumb that help you cut problems down to a manageable size. You read review articles, which pull together the key findings in a particular field over the last few years and summarize them in a few pages. You use RSS feeds and other tools that give you a heads-up when something new comes out. And you use one of the most invaluable tools in the researcher's toolbox when it comes to spread of information. It's fast, it's cheap, it's efficient—though that doesn't mean it's always right.

It's coffee.

It doesn't *have* to be coffee; tea, beer, or fruit juice will do the job just as well. The key point is that scientists from different disciplines often come together over a cup of coffee, exchanging information informally, off the record, and in a setting where you can ask questions that you'd never, ever put in a letter or an email. Questions like, How much can I trust those findings that the Innsmouth team reported last week? Questions like, How good is So-and-so when it comes to the quality of his data?

The word over the cup of coffee, in case you're wondering, was that Toth was a conscientious researcher who published good stuff. This was all the more reason to accept that he had beaten us to the finish line, and with a method that was better than ours. I had no reason to doubt what we'd been told, even though I hadn't read the paper.

Time passed, and I had almost forgotten about those cones of percussion. Then, one day over a decade later, I had a change of heart when I met a young master's student who was interested in handedness. Our method wasn't as good as Toth's had sounded, but it would work on flakes where Toth's wouldn't work, and it would probably be of some use to *someone*. So I offered Maureen Mullane a deal. I was too busy to catch up on the literature about handedness that had been published since Toth's study. She, however, *did* have the time. My ex-wife was happy for us to use the data; she was no longer working in flint research and was happy to have the data be of use to a young researcher. Maureen and I would coauthor the paper, crediting my ex-wife for her work on the original study.

Maureen started wading through the articles on inferring handedness from flint artifacts, including the Toth article. That's when she came back with a bombshell. It turns out Toth's method *wasn't* 100 percent accurate. In fact, it didn't appear to be as accurate as the results from our cone of percussion study. We hadn't been scooped after all.

We wrote a paper that ended up in *Laterality,* one of the top journals on human handedness. Maureen and I put the paper on our CVs and went our separate ways. We've stayed in touch, but I've seen her in person only once since then. The paper appeared in print, and I occasionally checked to see if anyone was citing it in research, but nobody was. You learn to live with that sort of lackluster response, if you have any sense and you want to survive in research. But in our case, there was a happy postscript: a bright PhD student named Natalie Uomini had picked up our paper and run with it, checking the cone of percussion method on a much larger sample size and confirming that, yes, it *was* as good as we had hoped. After getting her doctorate, she used that method, and others, in a study which concluded that Neanderthals were predominantly right-handed, in much the same proportion as modern humans.

It was a question that researchers had wanted to answer for decades. It was a surprising answer. Although Neanderthals were very closely related to modern humans, maybe close enough to interbreed with the first modern humans to leave Africa, they were very different in some key ways. They don't appear to have had art, whereas early modern humans were skilled artists. They were extremely conservative in their toolkit, with hardly any technological developments over tens of millennia. One simple explanation for those differences would be that they didn't have a sophisticated language. And if human language is strongly associated with human handedness, then it would tie the loose ends neatly together if Neanderthals didn't have handedness in the same way we do—for instance, if half of them were left-handed and half were right-handed, like most nonhuman primates, or if the majority of them were left-handed. But that wasn't the case. They were predominantly right-handed, just like us. So her findings answered one question and raised a batch of new questions, like whether the association between handedness and language might just be a coincidence, or whether the Neanderthals might have had a language as complex as human languages.

Nearly two decades had passed since I first started that work, and we finally had an answer to one of the big questions about early humans. If I hadn't dropped the ball for years, we probably would have had that answer much sooner. It's an example of one of the most unexpected findings that emerged from research into expertise:

LESSON 7
Experts make more mistakes than novices.

That sounds unlikely, but it makes perfect sense when you realize that experts tackling a difficult problem know more methods of tackling it than novices and will work through each method until they find the right one. Those aren't real getting-it-wrong mistakes; they're more like calibration tests.

My mistake about the Toth paper wasn't one of those, unfortunately. Still, I got there in the end, and I learned some invaluable lessons. That experience also brought home a key point. A hundred years of archaeologists had missed the significance of something so commonplace that it has been taught to freshman archaeology students down the years in Flint Tools 101 and its older equivalents. Now, when I'm talking to medical researchers or cold case detectives or others working in fields that seriously affect our lives, I wonder whether there's something in their field that's the equivalent of those skewed cones of percussion, something so familiar that nobody thinks to pay it any real attention, but which could transform that field.

Returning to handedness, further insights came later, through using other tools from the Verifier toolkit for representing the evidence. For example, instead of simply distinguishing between left-handedness and right-handedness, you could chart handedness on scales, the way Zoe Szymanski worked with the concepts of attractiveness and unattractiveness (figure 18).

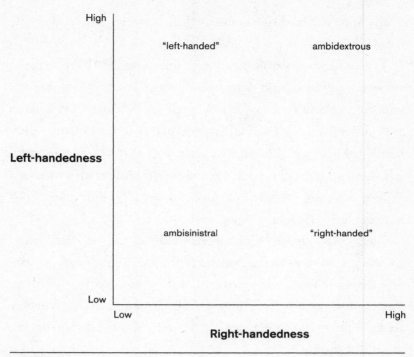

FIGURE 18

So, if you're skillful with your right hand and clumsy with your left, then you're strongly right-handed, and vice versa for people who are strongly left-handed. However, this chart also distinguishes between people who are equally skillful with both hands and people who are equally clumsy with both hands. In the research literature it had already been noted that a lot of people described as ambidextrous were in fact not equally skillful with either hand; instead, they were equally clumsy with either hand.

That alone would have been a useful contribution to the way people viewed the field. There's a further step beyond that, though. If you look more closely at people's skills and hand preferences, you start noticing some activities that don't fit neatly into the traditional distinction between left- and right-handedness.

Take calligraphy and pub darts, for instance. If I'm writing something in longhand, I'm strongly right-handed; I can't write anything clearly or comfortably with my left hand. Pub darts are a different story. I'm a fairly good player when I throw right-handed. However, I'm also a reasonable player when I throw left-handed, though it comes less naturally. And with a pick or a shovel, as a legacy of my field archaeology days, I'm fluent both left-handed and right-handed.

These obvious variations in handedness can be represented by using a number of diagrams like the one above, with a separate diagram for each activity. A further twist is that you can start looking at which activities show similar patterns: whether all the "throwing something" activities cluster together on the chart, for instance.

That sort of visual representation gives you a big stack of rich new questions to ask, with practical implications for training and job selection, as well as for academic research into the conditions associated with handedness. If you're trying to look for what might be associated with "handedness" and you're using a sophisticated, powerful set of definitions of "handedness," you're a lot more likely to get key insights than if you're just using the blunt instrument of "left-handed/right-handed." You could say the same for important research questions in the worlds of marketing, medicine, or manufacturing.

The better your representations, the better your definitions. The better your definitions, the better your questions and the better your answers. The better your answers, the better your chances of solving your problems, and of fame, or fortune, or whatever you're seeking.

The way you think about a problem makes a big difference

in the way you tackle it, and the way you tackle it makes a big difference to your chances of getting the outcome you want. Often the truth of a situation—success or failure—is in front of you, but for some reason you just don't see it.

If that sounds too abstract, here's an example that's as real and physical as it gets. It's from military history, and it's as clean and dramatic as anyone could ask for, showing how blind spots can lead to misperceptions that are quite literally fatal.

Ambushed on the Field of Blood

About two thousand years ago, a big Roman army was up against a much smaller enemy army. The Romans are legendary for being tough, organized, professional soldiers who knew everything there was to know about logistics and tactics, and all the little, low-level things that make a difference in hand-to-hand combat. This Roman army was up against a ragtag collection of warriors from across the known world: southern Italian cities in revolt against Rome fighting alongside mercenaries from Celtic Gaul, Spain, and even North Africa. It was a classic recipe for chaos, with troops who didn't speak the same language and who varied greatly in ability. They were fighting on a flat, open plain, which meant the Romans didn't need to worry about being ambushed—which was the only tactic that had a real chance against such odds.

Because the Romans had fought this sort of battle countless times, they could confidently predict the next stages. The Celtic cavalry would wipe out the Roman cavalry in the first few minutes of the battle, because that was what Celtic cavalry always did to Roman cavalry, and the victorious Celtic cavalry would then head off to plunder the Roman baggage train, because that was what victorious cavalry did after a successful charge; however, that wouldn't matter, because the Roman infantry would push forward through the center of the enemy army. After that, the process was simple: the Romans would break the smaller army in two, and then wipe out each half in turn. Divide and conquer—that's just what the Romans saw starting to unfold.

The Celtic cavalry wiped the Roman cavalry off the field within minutes in a bloody, brutal strike, and then rode off in pursuit of the fleeing survivors. The Roman general mentally moved those pieces off the chessboard and focused on punching through the enemy infantry. Slowly but steadily, the Roman army pushed the enemy center back until it was bowed deeply behind the flanks, on the edge of breaking, just as the Roman general had planned.

And then, in the middle of that open, featureless plain without hiding places, the enemy general sprang his ambush. His cavalry, which the Romans had written off as busily looting elsewhere, came storming back into the battle, and the Romans suddenly realized that, yes, you *could* view the battlefield as the Romans having pushed the enemy line back in a deep bow that was near to breaking. The trouble was, you could *also* view it as the Romans having got themselves surrounded on three sides, with their only way out about to be cut off by the returning enemy cavalry.

It was one of the worst "oh, shit" moments in military history.

The Romans were up against Hannibal, one of the greatest tacticians who ever lived. From the start he had planned to ambush the Romans with an army that was in full view on a featureless plain. His deception wasn't in a fold of the landscape or a river mist that concealed his troops till the critical moment. Instead, he had hidden the ambush in the Romans' own minds, setting up a scene that was deceptively familiar to them—the cavalry charging off in their usual wild way, the weaker line being pushed back—so that everything the Romans saw reinforced their belief that they were winning, until that last-minute switch when they realized there was a totally different way of viewing the situation.

That famous battle was at Cannae, in southern Italy. It still looks much the same now as it did then—a broad, flat plain dotted with olive trees under the hot Italian sun. If you visit it, the locals will point out to you the Campo di Sangue, the Field of Blood, where sixty thousand men were stabbed and hacked to death between dawn and sunset twenty-two centuries ago.

Trainee military officers around the world are still taught about that battle, and not just for historical reasons. Norman Schwarzkopf, the American general in charge of Operation Desert Storm, studied the Battle of Cannae in detail over the years and chose a modified version of the same tactic for a campaign over two thousand years after Hannibal, because it was that good.

Hannibal was the first general in recorded history to use the tactic now known as double envelopment. He also introduced another new concept to the military lexicon: the battle of annihilation. The soldiers he surrounded that day on three, then all four, sides were slaughtered down to the last man. It was arguably the highest death toll from a single battle in recorded history—worse than the first day of the Somme, far worse than Gettysburg or any of Napoleon's battles. More men were lost on that one day than the entire American death toll in Vietnam.

And it worked because of Hannibal's understanding of how his enemy thought. The Romans were defeated, not by something on the field of battle, but by what was lodged in their minds.

Birth of Verifier

What happened that day to the Romans is painfully familiar to many of us in business, science, and industry. The spreadsheets, reports, and data in front of us tell a story that looks reassuringly familiar. But then, if you look at the same evidence from a different angle, another interpretation is more likely and a lot less cozy. The tricky bit is knowing what that angle is.

If you talked to record company executives in 1999, they would have told you that they were in the throes of a familiar battle with pirates who were ripping off their digital content. But they reckoned that the suits had the situation well in hand. They had everything on their side: the lawyers, the money, the law. It was only later, when the music business changed forever, that those executives could look back and see the truth: the digital revolution had begun, and many of them were about to face redundancy.

We always believe that we're looking at the problem correctly. The Roman soldiers did. That century of archaeologists did when they looked right past the clues on those stone tools in their collections.

I had blundered too, but that mistake had moved me forward. It had given me firsthand experience of seeing something that the experts in a specialist field had all missed, and had given me the confidence to try looking for bigger mistakes in other fields. With the handedness study, I simply happened to have a good idea and got lucky. Now I was wondering whether there was some way of systematically looking for those missed avenues.

It was an ambitious idea, but it was far from impossible. There were plenty of previous examples of a researcher changing a field of science by bringing in a single new method. When the

Princeton psychologist Philip Johnson-Laird introduced game theory from mathematics into evolutionary ecology, it transformed the field overnight by giving researchers an immensely powerful new tool for analyzing their domain. Game theory wasn't new—it had been around for centuries—but spotting its implications for evolutionary ecology, the scientific study of how behaviors, adaptations, and mutations affect the survival chances of an individual creature or an entire species, was a true breakthrough.

On my long commute into work at Middlesex University, where I was a senior lecturer in computer science, I did a lot of thinking about this idea. I was at the stage where I knew some chunks of the puzzle, but I still didn't know how those chunks fit together.

One chunk involved what comes upstream of formal logic. The formal logicians, people in the tradition of Aristotle and the medieval logicians, did a great job of modeling error within a given case, such as the classic "logic fail" example: "All dogs have four legs; that animal has four legs; therefore that animal is a dog." The problem with this approach was that it was wide open to what's known as GIGO—Garbage In, Garbage Out.

If you gave formal logicians a description of the reasoning behind the design of a ship's hull stress monitoring system, they would do a great job of analyzing the reasoning in that description, but they wouldn't know whether the initial description was a complete, correct description of reality.

So you needed some way of checking whether what was going into their logic was trustworthy. If some material going into their logic was from written sources, like handbooks, technical specs, and journal papers, you needed to understand the realities of the worlds that formed the basis of those written sources. That would take you into the research literature on the sociology of science. You also needed to know about human errors and biases, so you could watch out for them in both the written sources and what you gleaned from human experts. Lastly, you needed to know about how best to represent all that knowledge, because a different visual representation as simple as the chart I showed you for handedness can provide profound new ways of tackling a problem.

Those were good, solid chunks, and I had experience using all of them. I could see how to fit them together to investigate some aspects of errors in academic reasoning. Using those approaches in combination would be a brute—there were so many concepts in there from so many different literatures. It would take years to master them all if you were starting from scratch. But that brute, in the end, could be very powerful. The problem was that there were still chunks missing. I knew they were missing, but I couldn't put my finger on what they were.

One major chunk had been right in front of me for months, and I'd been too close to realize its significance. When the answer finally popped into my head, I knew immediately how significant it was, and I wanted to tell someone about it as soon as I could. The person who had that idea dumped onto them was a first-year PhD student named Jo Hyde.

Jo was working on a PhD in computer science supervised by my colleague Ann Blandford and me. If you've ever watched early episodes of *The X Files,* then imagine Jo as a trainee FBI agent being mentored by Mulder and Scully. Jo was a determined individual. Right at the start, we nailed down an explicit working agreement. Ann, as lead supervisor, would oversee the formal, academic side of Jo's PhD—the stuff relating to the topic of Jo's thesis on interface design. I would coach the informal side—the low-level craft skills of doing a PhD. I would show her how to do data collection and help her build the sort of career she wanted.

Jo was particularly good with deep structure. Linguists make a useful distinction between the surface structure—what something *looks* like on the surface—and deep structure, which is what's going on underneath. If you strip away the surface detail of a movie plot, for instance, you'll often find that it has the same underlying deep structure as another movie, a well-known novel, or a play. Johnson-Laird had recognized that the deep structure of some key problems in evolutionary ecology was the same as the deep structure of some already solved problems in game theory.

People who have the knack of spotting underlying deep structures are extremely valuable to all fields, not just their own. If you're good at seeing deep structure, then you've got a good chance of being able to spot ready-made solutions to problems that everyone else is having trouble with. It also makes you first choice as a sounding board for anyone who's wrestling with a big idea or a complex problem.

My big idea involved spotting a missing piece of the jigsaw. I had been grappling with the GIGO problem, and I'd been thinking about the problems involved in digging the facts out of the literature and representing them in the right way, ready for formal logic. I had completely overlooked the issue of getting facts directly out of the experts themselves. I had completely forgotten to consider my own work with Neil. As soon as I did, the implications were obvious.

If you combined the ACRE framework with the other pieces of the jigsaw that I had been assembling, you had pretty much all the pieces you needed to solve the first part of the GIGO problem. You had a toolkit that would let you check everything that was going into the system, so that the logicians would be working with real-life raw material that was complete and dependable. The formal logic part was something I hadn't even started to think about because I was sure it would be a nightmare, but at least I had put together a solution for the first half of the puzzle.

I ran into Jo outside the cafeteria. We stood by one of the atrium pillars, out of the flight path of the hordes of students and staff, while I trotted out my new idea. She listened, quickly grasped the deep structure of what I was saying, spotted what was missing from my idea, and hit me with a single sentence that supplied the rest of the puzzle.

She pointed out that I could combine my new idea with a faceted taxonomy of formal logics. That wouldn't mean anything to most people. For me, it was one of those times where your life changes in seconds.

With the idea that I had described to her, I'd worked out how to handle the "Garbage In" part of the equation. I hadn't even started on the problem of how to process what came out. For that problem, I knew formal logic would be necessary, but I hadn't worked out how to do that.

The problem is that there's not just one type of logic; there are lots of types, and we would need to work out which ones to use for a given problem. But Jo's insight solved that problem. Her insight was that you simply treated the types of logic the same way that Neil and I had treated the different elicitation methods. Our ACRE framework said: for *this* type of knowledge, use *this* elicitation method to get the knowledge out. And now, Jo was saying, for *this* type of information, use *this* type of logic to verify the expert's work. That's what her killer sentence was saying.

It was simple and brilliant. It was the missing piece. We now had a complete, integrated method that could be a ground-breaking new way of checking for errors in expert reasoning. The obvious next step was to try it out on a real-world problem, something that was defying all attempts at a solution. And we knew exactly which problem we wanted to tackle.

6

The Voynich Manuscript

WHEN JO HYDE AND I PUT THE PIECES OF OUR method together in that fateful conversation outside the cafeteria, our immediate target of choice was Alzheimer's. We were fairly sure that our method, which didn't even have a name yet, would work best for problems where researchers already had all the information they needed, but they hadn't been able to figure out how to put the pieces of their jigsaw together. There was no point in applying it to a problem where the experts didn't yet have all the necessary information; the method was designed for cases where research was probably being held up by faulty assumptions or faulty reasoning.

We were well aware of how arrogant this might sound, but remember one of our first rules about experts: their expertise is limited to the areas in which they're experts. The biochemists, the neurophysiologists, and all the other experts working on Alzheimer's knew far more about their own fields than we would ever know. However, if the missing piece of the puzzle involved a flaw in human reasoning or the way information about a key feature of Alzheimer's was being represented, we could help them. If we could identify and check the core assumptions and reasoning the experts were using, we had a real chance of spotting solutions everyone else had missed. There were a lot of different disciplines working together on Alzheimer's, and that meant that there was a good chance of misunderstandings where disciplines overlapped. It was a promising target. There was also the consideration that if you think you have even a small chance of cracking a problem as grim as Alzheimer's, the choice is pretty clear. You've got to give it your best shot, even if the odds are heavily against you; walking away without trying isn't really an option.

But we were not planning to tackle it by ourselves, because we knew how big the problem was. We were sure that we would have to work in collaboration with Alzheimer's researchers.

We made a simple pitch to one of the big pharmaceutical companies: "We have an unproven method that is likely to be time-consuming and has maybe a one percent chance of success. But if it does pay off, the payoff will be huge. And because people make the same sorts of errors in whatever area they're working in, you can use the same method in any other area you want."

Their response was sensible but depressing: "Go out and crack an unsolvable problem to show us your method works; then we'll consider applying it to one of our medical problems."

We had been half-expecting that response but had hoped for a bit of luck. That hadn't happened. We reckoned that we'd need a lot of time and funding to get a successful demonstration of our method. To get that, we would first need a successful demonstration of our method, which would require having the time and funding. . . .

It looked like we were locked in a catch–22. Within a month of its invention, the method we were now calling Verifier would be stuck on the back burner unless we could find an impossible problem that we could solve for pennies. And then by chance I found what we wanted, though without realizing its significance until much later.

The Voynich Manuscript

Some people do crosswords in their free time; I prefer to read about real-life mysteries. I'm always interested in puzzles that haven't been solved. Some mysteries are tantalizing: was Plato's legend of Atlantis based on the eruption of the volcanic Greek island of Santorini? Does the 1967 Patterson-Gimlin film really show a Bigfoot with features that nobody could fake at the time? Most, though, are depressingly overhyped.

Then one day I stumbled across a mystery in a league of its own. The more I read about it, the more solid the evidence became, and the deeper the mystery.

This is no grainy film or unverifiable eyewitness story. It's a solid artifact. It has often been called the most mysterious book in the world; nobody has ever been able to understand its contents because it's in a strange writing, illustrated with hand-drawn pictures that are a mixture of the mundane and the bizarre. There's no other book in the world that looks anything like it.

If you want to see the Voynich Manuscript in person, it's in the Beinecke Library at Yale University. It's a real, solid book the size of a large paperback.

When Wilfrid Voynich first described the manuscript he had

discovered, he gave a vague account of where he found it. He later explained that he was deliberately keeping the location secret because he had not been able to examine all the books that were being sold and wanted to see the others before telling everyone his source. From a commercial point of view, it was a perfectly sensible explanation, but it was to raise suspicions that the manuscript might be a hoax produced by Voynich himself aided by his counterfeiting contacts.

Serious researchers now acknowledge that the documentation trail makes it clear that Voynich made active efforts throughout his life to have the manuscript deciphered, well beyond the token efforts needed to add credibility to a hoax. Voynich found the manuscript in a box of books offered for sale by the Jesuits at the seminary in the Villa Mandragone, just outside Rome. Most of the books were ordinary, but one immediately caught his attention. At first sight, it looked inconspicuous. It was about the size of a thick paperback novel, written on medium-quality vellum.

The writing is clear, but it's the writing of a working document, not a book meant for show. In a professionally produced manuscript of the late Middle Ages, the lines would be neatly laid out on the page, the edges of the text aligned to lines faintly drawn in advance. The text of the Voynich Manuscript was written without any of these preparations, its lines wavering, evidently aligned by eye rather than along ruled lines. The illustrations are prolific and fluently drawn, without visible corrections, but like the text, they are not of the quality that would be found in a professionally produced manuscript intended for display.

It looks, in the words of one researcher, like a perfectly ordinary book—perhaps the notebook of an herbalist or alchemist.

What struck Voynich, though, was summed up in the full quotation by the same researcher: it looks like a perfectly ordinary book that happens to have dropped into this world from a parallel universe.

The illustrations look ordinary, too, at first glance—a water lily, or a heartsease, the sort of plants that would be expected in a late medieval herbal. At second glance, though, most of the illustrations are unsettlingly alien: bizarre plants, clearly nothing found on this earth; naked women bathing in strange channels that look in some ways like plumbing; zodiacal images with subtle differences from the zodiacs of the time.

The text also looks familiar at first glance, particularly to anyone who has tangled with the varied handwriting of medieval manuscripts; it has a lot of similarities to various traditional European scripts. At closer inspection, this appearance of familiarity dissolves: the script also has many unique characters. Usually manuscripts make more sense as the reader becomes familiar with the script and realizes that this odd-looking character is actually an *f* and that other one a *d*. With the Voynich manuscript, the more you look at the script, the more you become aware that you can't read a word of it.

Voynich immediately realized the significance of his find. The manuscript appeared to be a late medieval book, probably written by an alchemist or herbalist as a personal notebook, but entirely in code. It was a plausible idea—alchemists often used codes to conceal their findings and included bizarre allegorical illustrations that were not intended to be taken literally. What was remarkable was the extent of the document. If genuine, it would be the longest cipher text of its type ever discovered, as well as the most prolific example of allegorical illustrations. There were partial precedents: some entire books in cipher text had been produced in Italy in the late fifteenth century. Other books included alchemical illustrations in the same spirit as those in the manuscript—a bush growing out of the back of a carp, for instance. There were few datable features in the illustrations or handwriting, but those few did point toward northern Italy and the late fifteenth century as the origin, which was consistent with the apparent content and nature of the manuscript.

Voynich bought the book, returned home, and immediately set about having it deciphered, contacting friends in the cryptographic community for help.

. . .

Voynich had become rich by selling old, rare books. He knew within seconds of opening this particular book that it was unique. There are a few old handwritten texts with odd illustrations and a few in code, but the illustrations in the Voynich Manuscript go beyond odd into downright surreal, and there's no other cipher text anywhere near this size—it's well over two hundred pages long. It was the find of a lifetime.

Deciphering the text should have been easy. It looked as if it dated from around the late fifteenth century, in terms of the style of handwriting and illustration. The late fifteenth century was a golden age of codes, particularly in Italy. The Italians had two institutions that needed codes: warring cities and the newly invented banking system.

In 1912, when Voynich found the manuscript, cryptography was still important for the same reasons: war was looming, and international trade depended on codes. There were differences in scale: the looming conflict was World War I, and international trade was no longer confined to Western Europe; it included entire continents that hadn't even been discovered in the days of the early Italian codes. Cryptography had moved on, too. The early codes were known, taught to first-year undergraduates in history of cryptography classes, and now long abandoned for serious code work—their shortcomings had been exposed centuries before, and modern codes were more sophisticated. If the Voynich manuscript contained a fifteenth-century code, a good modern cryptographer could expect to crack it in a few days at most. But they didn't.

When the Internet arrived, it depended on secure codes. Cryptography became even more important. By this time, codes were massively more sophisticated than their fifteenth-century ancestors, but even the best modern codes were still being cracked within a few months or years at most. But the Voynich Manuscript remained undeciphered ninety years after its discovery.

By this point, it had become a legendary challenge for cryptographers. Whoever cracked it would demonstrate world-class mastery of their field. There was also the consideration that the code might hold the key to an entire new generation of ultrasecure codes; if so, whoever knew its secret could become phenomenally wealthy overnight. However, that all depended on the assumption that the manuscript did contain a code.

Over the years, suspicion had grown among researchers that the manuscript might not contain a code after all. One obvious possibility was that it was a hoax, containing only meaningless gibberish; another was that it was an uncoded document written in an unidentified language. However, there were serious objections to both explanations. The text in the manuscript shows complex regularities, so complex that it was hard to imagine anyone being able to produce and sustain them throughout over two hundred pages of text. Some statistical features of the text involved factors that weren't discovered until centuries after the manuscript's likely date of creation.

Those same regularities were also hard to reconcile with an unidentified language: no known language was anything like the text of the manuscript. If it was a language, it was the strangest language ever discovered; if it was a hoax, it appeared to be the most complex hoax ever invented, and nobody could work out how it could have been done.

. . .

You can tell a surprising amount about a mysterious manuscript without being able to understand a word of it, if you know what to look for. I'll show you how to do it, starting with the page usually known as the "sunflower" page (figure 19).

For starters, we can make a good guess as to what the page is about. The large picture of a plant is a strong hint: this is probably a page about a plant. Most pages in the first half of the manuscript follow the same pattern: a picture of a single plant, taking up most of the page, with a few lines of text fitted around the picture. That may not sound like much, but it's actually a very, very good start. The next couple of examples show why. They're both modern codes.

```
75628  28591  62916  48164  91748  58464  74748  28483  81638  18174

74826  26475  83828  49175  74658  37575  75936  36565  81638  17585

75756  46282  92857  46382  75748  38165  81848  56485  64858  56382

72628  36281  81728  16463  75828  16483  63828  58163  63630  47481

91918  46385  84656  48565  62946  26285  91859  17491  72756  46575

71658  36264  74818  28462  82649  18193  65626  48484  91838  57491

81657  27483  83858  28364  62726  26562  83759  27263  82827  27283

82858  47582  81837  28462  82837  58164  75748  58162  92000
```

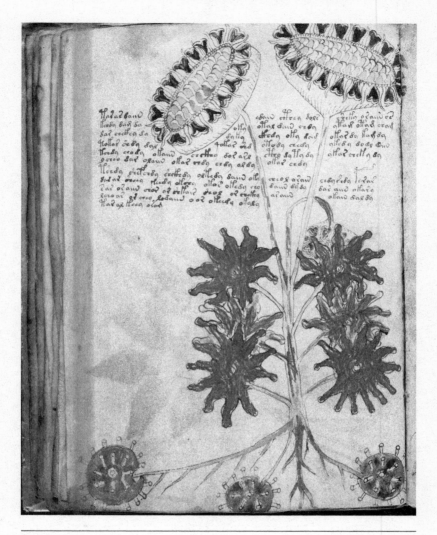

FIGURE 19

The first example is the D'Agapeyeff Cipher from the 1930s. What's it about? I have no idea. Nobody else has any idea either. Alexander D'Agapeyeff, the mapmaker and author who conceived it, forgot what coding system he had used. There's no clue about the content of the text. Also, the text has been broken into chunks of five characters, so we can't tell where one word ends and the next begins. That's a standard feature of modern cryptography which makes life harder for would-be code breakers. However, we do at least get one useful clue from its structure: the last line is shorter than the others and is left-justified, which tells us that the text should be read left to right, starting with the top line.

The next example is from a cipher text that I produced, called the Penitentia Manuscript (figure 20).

Which direction should you read it in? There's no clue. The symbols form a complete square. All the other pages of the Penitentia do the same, and they're arranged in a square of four pages by four. There's no way of telling whether you're sup-

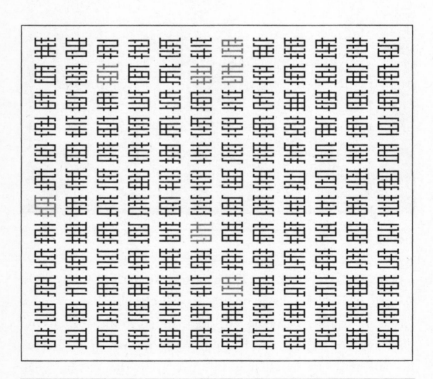

FIGURE 20

posed to read it top to bottom by columns, or left to right, or right to left.

With the Voynich Manuscript it looks as if would-be readers have been luckier. It's written left to right, just like most Western European languages. That's a good start, and having pictures on the pages is even better. If you know that a page is probably about a plant, there's a fair chance that the page will contain words like *leaf* or *flower* and probably the name of the plant itself. So what else can we tell just by looking at a page of the manuscript?

Looking at the letters and the words, you've already identified something significant (figure 21). It's written using an alphabet and separate words. That may not sound like much, but it's actually a promising break. A lot of writing systems don't use alphabets; they may have each character represent a syllable, or even an entire word. Also, a lot of old writing systems didn't bother to separate words clearly, which adds another layer of complexity.

The alphabet itself also looks like a lucky break. Quite a few characters look familiar—there are characters that look like *o* and *a* and *8* and *9,* and there's another that looks a bit like an *n.* The tall, flourishing characters may not look familiar to most people, but to experts in handwriting styles they look similar to a style that was common briefly during the fifteenth century, a style known as the "humanist hand." That gives us a fair idea of when the manuscript was written.

When you look at the pages, you start seeing groups of characters that are repeated in different words. One example is a pair of characters that looks like *40.* This pair often appears at the start of a word but hardly ever occurs anywhere else. This implies we're dealing with a straightforward language rather

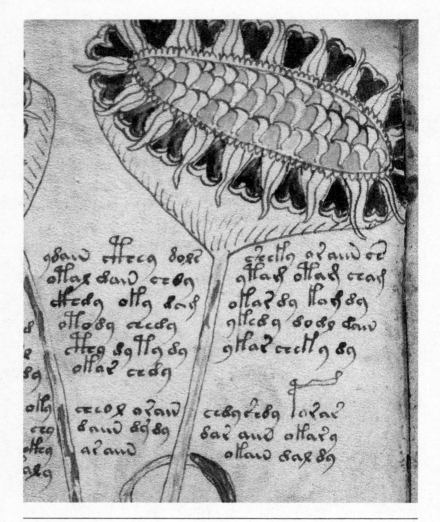

FIGURE 21

than a code. Good news. You get almost exactly the same pattern with *qu* in most Western European languages—the *u* often occurs on its own, but the *q* almost always occurs in a pair with *u*, and the pair never occurs at the end of a word.

Because regularities like this make it easy to narrow down possibilities, code makers hate them. One of the first things you do in a code is to hide this sort of regularity, whether by rearranging the order of the letters or in some other way. If you see separate words in a text, and those words have regular groups of characters within them, then you're either dealing with a perfectly ordinary language using a script you don't know or a very simple code. You can try this with the Voynich Manuscript.

And that's where the good news comes to an end.

. . .

It all starts to go wrong when you look for the plant names. Names are great if you're trying to make sense of an undeciphered document, whether as a linguist or as a code breaker; they're a really good way in. That's how Egyptian hieroglyphs were cracked.

Let's start with an easy example. Suppose you see a page with a picture of a lily, and the first word on that page is four letters long, with the same character as the first and third letters—for instance, *8089*. That's the same pattern as *lily* in English, so maybe you're dealing with a text that's in English, where *8* stands for *l* and *0* stands for *i* and *9* stands for *y*.

A lot of English words end with *ly;* a lot of words in the manuscript end with *89*. That looks promising. However, a lot of words in the manuscript begin with *89,* but hardly any English words begin with *ly*. That's the principle.

Each of the plant pages in the manuscript begins with a word unique to that page, suggesting that the word is the plant's name. If you try to fit those words onto the English name for the recognizable plants (a water lily, for instance), you soon find that the text in the manuscript can't be made to fit with English; it must be something else.

"Something else" may sound like bad news, but it's actually not. There are thousands of languages on the planet, but most of them fall into broader families, where a lot of the features and vocabulary are similar across family members. English is a member of the Germanic family of languages; the English word *hound* is similar to the German *Hund* and the Dutch *hond*. So you don't need to check the text of the manuscript against every language in the world; you just need to check it against the most likely language families. The number of language families is fairly small—dozens, rather than thousands. When you start looking at other languages and other language families, you will soon find that the manuscript wasn't written in the usual scholarly language of the time, Latin, or in Greek, or in any of the other major European languages.

A further complication is that medieval scholars were often secretive about their findings and invented cryptic names for things to conceal their knowledge from potential rivals. A code breaker can sidestep that complication by looking at low-level language features that encoders wouldn't bother to change.

Smaller words are a good place to start. The most common words in a language are usually also the shortest ones, like *a* and *the* in English. If you're lucky, some of the words in your manuscript will turn out to be close to short, common words in a known language or language family. The word *in* is used with the same meaning in English, German, and Latin, for instance—three languages in two language families. There's an obvious risk of random coincidences, but if there's a systematic pattern of correspondences between words in your manuscript and words in a known language, you're probably dealing with a manuscript in a related language, so you'll be able to work out what the text in your manuscript means. There are plenty of common short words in the Voynich Manuscript, but when you try this approach on them, you get nowhere. Nobody has ever found a convincing set of correspondences between them and any known language.

In a linguistics course, one of the things they teach you is the giveaway features of different languages and language families. Thus, if you encounter an unknown language, you can swiftly work out what it might be related to. Do the middle and end of the word remain constant, while the first couple of letters change? If so, you might be dealing with an African Bantu language, where you form plurals by changing the start of the word instead of the end. If the beginning and middle of the word remain the same, but there are a lot of different endings, you might be dealing with one of the many Indo-European languages, like English, German, and Latin. If there are restrictions on which vowels occur together within the same word, you're dealing with a language that has vowel harmony, like Turkish. And so on.

When you look at the text in the Voynich Manuscript this way, you soon eliminate all the obvious candidates. Its features don't fit with the Indo-European language family (most of the languages in Europe) or the Semitic language family (such as Arabic and Hebrew). That's already odd. The vast majority of texts from fifteenth-century Europe are in either an Indo-European or a Semitic language. Its features also don't fit with Bantu (a huge number of African languages), Finno-Ugric (languages spoken in Hungary, Finland, and Estonia, among other nations), or Turkic (languages spoken in Turkey, Siberia, and other parts of the world).

There are a handful of languages in Europe that are unrelated to any other known language: Basque is the most famous example, a solitary survivor spoken in the Basque region of northern Spain and southwestern France, whose ancient relatives were wiped out and replaced by Indo-European languages thousands of years ago. But when you try to map the features of Voynichese onto them, it doesn't work. If the manuscript is written in a language, it is a very exotic one. Or it is no language at all, just mere gibberish. But there are serious problems with that explanation, too.

.   .   .

Real languages have a lot of regularities. For instance, pretty much every real language includes small groups of words that tend to occur in a predictable cluster. In English, if a sentence has the phrase "the more" early on, it will almost certainly have either "the more" or "the less" later on. That's another way into an unknown language: if you can spot a few of these clusters, you have a good chance of working out what language you're dealing with. The text in the Voynich Manuscript doesn't have any regular pattern in word distribution. However, there are regularities that do occur both in real languages and in the manuscript, and which are very difficult to fake. One example is the distribution of common syllables. In the manuscript, the syllable *40* occurs frequently at the start of words, but hardly ever occurs at the end. Some letters occur either as singletons or in pairs (like *e* and *ee* in English), but other letters only occur as singletons (like modern English having *u* as a common letter, but hardly ever *uu*). It's easy to imagine someone making up a fake language that has a few of these features, but the sheer number of these regularities and the way they're maintained across so many pages of text make a hoax much harder to imagine.

One huge question for the hoax hypothesis involves statistics. If you take a book written in a real language and plot how often each word occurs, you get a characteristic curve that is very similar across all the real languages that have been tested. A few words, like *the,* are very common, most words are moderately common, and some only occur once or twice. If you do this for the words in the manuscript, you get a curve that is similar to the curve for real languages. It's hard to imagine how anyone in the fifteenth century could have faked that, even if they had known about it. For this reason, a few people wondered whether it might have been faked either by Voynich himself or by someone who planted the manuscript.

There are various other statistical tests you can apply to texts; on all of those, the text from the manuscript came in close to the values for real languages. Those statistical tests also confirm something else that researchers suspected from the start: not only is the manuscript written in at least two different styles of handwriting, but it's also written in two different "dialects," which have subtly different textual characteristics. That level of sustained complexity and regularity is hard to explain as part of a hoax, or as someone speaking in tongues, or any of the other mechanisms that have been suggested for producing meaningless text. If it is a hoax, it is an astonishingly good one.

There have been only a couple of known hoaxes that were anywhere near the same league. One is the languages channeled by "Helene Smith," a patient of the Swiss psychologist Théodore Flournoy in the nineteenth century. Smith claimed to be receiving messages from Martians. The other is the Enochian language channeled by the English adventurer and confidence artist supreme, Edward Kelley, in the sixteenth century. If you know what to look for, you can spot where these languages came from. The word order in Enochian sentences is the same as in Elizabethan English. As anyone who learns a foreign language knows, speaking another language isn't simply a case of translating word by word; you also have to learn the phrasing and structure of those foreign languages. Smith's Martian tongue had the same problem. If you scratched the surface of her Martian writing, you found that the deep structure was identical to the deep structure of Smith's native tongue, French. But the text in the Voynich manuscript doesn't look like it's based on English or French or, for that matter, any language in the world. It's unique; nothing like it has ever been found.

If the manuscript was written in an unidentified language, it was the strangest language ever found. The hoax explanation also had serious problems: nobody could see a way of producing a hoax with such regularities, even with computer-age technology, let alone fifteenth-century tools and skills. That left a code as the only plausible explanation, more or less by default.

Voynich had hoped to see the manuscript cracked within a few years. That didn't happen. Any known code from that period should have been cracked within days. That didn't happen. One possibility was that the manuscript was written in a super code that was centuries ahead of its time. Every year that passed, with every advance in cryptography and computing power, left that possibility looking less and less believable. But none of the other options looked any more believable: it was either an astonishingly strange language, or an astonishingly good hoax, or an astonishingly good code.

One obvious possibility is to try working in the other direction, from the list of potential perpetrators, and see whether that gives any insights into how the crime could have been committed. And when you do this, one name leaps right onto center stage.

Brightest of Angels

Edward Kelley had always been a prime suspect for hoaxing the Voynich Manuscript. He was a prime suspect for pretty much anything. Kelley walked into history on the eighth of March 1582. He vanished from it fourteen years later, climbing out of the window of the tower where he was imprisoned, using a rope made of bedclothes, and disappearing into the night. In the years between, according to contemporary records, he spoke with angels, transmuted base metals into gold, was a soldier of fortune, and became a baron.

We don't know anything for certain about his earlier life, before his first meeting with John Dee. There's a Kelley in the records who was probably him, and there are a few stories involving forgery and theft and necromancy. Some of those are probably true. We don't know what happened after he drugged his guards and escaped into the night. There's some evidence that he died from injuries during the escape, but there's also some evidence that he was alive and free years later.

Dr. John Dee, Kelley's master and victim, was no ordinary doctor. He was one of the greatest intellects of the sixteenth century, an authority on the classics, an advisor to Elizabeth I, an able administrator chosen by the queen to reorganize the navy's dockyards, and a practical mathematician who played an important role in reforming the calendar. His main goal, though, was to communicate with angels. He hoped Kelley would be the medium through whom that communication would be possible. That hope was the key to their fateful association and eventual tragedy.

The usual depiction of Edward Kelley's first meeting with Dr. John Dee is of a demonic Mephistopheles meeting a credulous Faust, but the reality is far more complex, and in some ways stranger than its fictional counterpart. It is easy to read that first meeting as a fraudster tricking his way into his victim's confidence, but that wasn't what happened. On that March day, Kelley had come to Dee's house because Dee had invited him, for reasons related to a radical new application of theology to pragmatic problems. Each step of Dee's reasoning was based on principles still believed today by devout Christians; his brilliant insight was in the way he had combined them. He knew what he was doing, and he had a shrewd idea of what he could expect from Kelley.

Dee wanted a shortcut to dependable knowledge, in an age of intellectual turmoil when the centuries-old classical texts could no longer be regarded as reliable authorities. It was an age when politics were driven by gunpowder and gold from the New World. Medieval scholarship was based on texts from classical antiquity, written more than a thousand years before; that scholarship offered no useful insights into sixteenth-century gun design or the mapping of the New World. A new scholarship was growing, with many false starts and blind alleys, and with few solid fixed points on which scholars could build. Dee had realized there was a way of jumping past this stage.

It was clear from the Bible that angels sometimes spoke directly to mortals. Dee's insight was that it might be possible to talk to the angels, not in prayer but in a rational discussion. Instead of praying to the angels like a student begging a tutor to intercede with a headmaster, Dee thought he could approach the angels like a student asking a tutor for information about his studies, thereby gaining access to divine knowledge. It was a bold concept based on solid theology—an important factor, since it might save Dee from accusations of sorcery and the very real risk of being executed, with all the cruelty of sixteenth-century justice. Its theology was solid; in fact, anyone questioning Dee's basic premises would themselves risk being accused of heresy, for denying that angels spoke to mortals or that angels heard mortals' prayers.

The obstacle stopping Dee was a practical one: he couldn't hear what the angels said, and it was clear from scripture that angels only spoke to mortals in special circumstances. Here, Dee's solution was based more on popular belief than on scripture. It was widely believed at the time that some people were particularly sensitive to supernatural forces; some of these, Dee thought, might be able to act as go-betweens in his communications with the angels. He had already tried this more than once, with disappointing results. The sensitives he employed—people called "skryers"—were generally expected to behave differently from normal people as a result of their special powers, acting more impulsive, moody, or difficult, much like the modern stereotype of the gifted, difficult artistic temperament. Those he had encountered fitted this expectation but produced only thin, weak accounts of the messages they claimed to have received. Kelley was to be another such attempt. He had been recommended to Dee by a mutual acquaintance, one Mr. Clerkson, who knew various skryers and other wanderers on the edges of society. Dee was expecting someone temperamental, unreliable, and possibly dishonest; that was a price he was prepared to pay, provided that Kelley could produce what Dee wanted. Dee was to find himself more correct on all counts than he had ever imagined.

The man who approached Dee's house that day was inauspicious in appearance: walking with a stick, in a long cloak with its hood pulled up. Legend says that the hood was up to conceal an ear cropped for counterfeiting, but there is no solid proof of that. Nothing is known for certain about Kelley's previous years; there are contemporary records of an Edward Kelley who was probably the same man, and there are memoirs written after the event whose descriptions are colored by knowledge of what was to happen. Those memoirs contain rumors that he was an unfrocked monk, a counterfeiter, a necromancer. In 1582 he was a skryer, and that was why he had been invited to meet Dee.

The invitation was, for Kelley, an important one, a chance to break out of the margins of society in which skryers existed. An indication of Kelley's position is that he introduced himself to Dee under the assumed name Talbot. Modern writers speculate that this was because he believed that the hint of a link to that family, famous at the time, would bring him some reflected status. Another speculation, from a contemporary of Kelley's, was that it was to conceal his whereabouts from a manhunter pursuing him for the theft of jewels. Possible thief, possible unfrocked monk, possible counterfeiter—this was a chance to leave his past behind.

Former clerics were nothing new to Dee—he was employing a former priest, as a skryer, when he invited Clerkson to bring Kelley for the initial meeting. Barnabas Saul, the former priest, disappeared the day before that meeting, after several months in Dee's household. What the reasons were for that disappearance, we can now only conjecture. Saul was the latest in a line of skryers that Dee had employed since at least 1579, and probably for several years previously. Their presence had not gone unnoticed: there had been rumors for years that Dee was communing with devils, and he had been exasperated by these rumors, unable to strike back effectively against them, reduced on one occasion to publishing a refutation of these rumors in the hope that it might set the record straight.

Against this backdrop, Kelley's murky past was nothing remarkable, and the initial meeting went well. Kelley and Clerkson soon returned, invited to dinner, and Kelley joined Dee's household as the new skryer, taking the place of the vanished Barnabas Saul. Kelley was now living in the household of one of the foremost intellectuals in sixteenth-century Europe, and his glittering rise to short-lived but dazzling fame had begun.

Kelley convinced Dee that he could indeed communicate with angels, and in the séances that followed, he dictated to Dee the lost language of Adam and of Adam's immediate descendants—Enochian, a language with an elaborate alphabet much favored today by a certain type of alternative scholar and by a certain type of tattoo parlor. It's a huge, elaborate language, and when you look at it through the eyes of a modern linguist, you can see it's nothing more than a glittering scam. Its internal word structure is incoherent; its word order maps one-for-one onto Elizabethan English. It's a hoax, but a brilliant one.

It was complicated enough to convince Dee. That set him up for the next part of the story, where Kelley announced that the angels wanted Dee to travel to the court of Rudolph II, Holy Roman emperor, the most powerful man in Europe, and reprimand him for his sins. Dee was convinced that these were the words of angels, and he reluctantly agreed. The two men traveled to Europe to meet with Rudolph. Dee's reputation was such that he actually got away with reprimanding the emperor, though it didn't endear him to Rudolph.

Then things went in a completely different direction. Kelley convinced Dee that the angels wanted the two men to swap wives for the night. Dee and his beautiful young wife were aghast, but the angels, via Kelley, were insistent, and what Dee called in his diary "the great cross-match" went ahead. His marriage was strained to the breaking point; it took years to patch it together again. A broken man, Dee returned with his family to England, leaving Kelley as a rising star famed for his ability to turn base material into gold, a pupil who had eclipsed his master and was now firmly ensconced in the court of the most powerful men in Europe.

Rudolph was an obsessive collector of anything strange. There's a story that he once bought a relic claimed to be "the skull of John the Baptist *as a young boy*." And the records show that he bought a book which is almost certainly the Voynich Manuscript, at just the time when Dee and Kelley, that notorious hoaxer and confidence trickster, were in Rudolph's court.

Kelley's career was spectacularly varied over the next few years—he became a baron, he became a soldier of fortune, and he was later imprisoned. Finally, in an episode straight out of fiction, Kelley drugged his guards, made a rope out of his bedclothes, and vanished out of his tower window into the night, returning to the murky half-world from which he had emerged those few short years before.

Fascinating though this story was, I didn't think that the Voynich Manuscript was a suitable test case for the Verifier Method. I didn't want to spend years reading up on cryptography, when everyone agreed that the content of the manuscript would at best contain the ramblings of a medieval alchemist. That just wouldn't give enough payoff for the time I expected it to take. I read about the manuscript intermittently from then on but didn't do any serious work on it until something happened that is all too rare for researchers: I unexpectedly had some spare time.

Hands-On Science

I was about to start a new job at Keele, working with some of the people I had met at that conference where Neil and I presented the ACRE framework, which spelled out which elicitation techniques would help you get a specific type of knowledge out of experts.

Keele has a beautiful campus, surrounded by beautiful countryside. My partner Sue and I had made the decision to move there because of its quality of life. The timing of the move meant that I had several months with a light workload, which left me with free evenings and weekends. I used that time to take a closer look at the Voynich Manuscript. Some aspects of the literature on the manuscript had been bugging me. Three in particular stood out.

One was the "it's too complex to be a hoax" argument. In fact, it's easy to produce complex outputs using simple processes. One of my long-term hobby goals in archaeology research is to make a pattern-welded sword. The blade of a pattern-welded sword has complex, semiregular patterns in the steel (hence the name). Modeling these patterns mathematically or via computer program would be a nightmare. However, producing the pattern welding is simple: you just take a couple of iron rods, heat them in a forge, twist them together, and beat them into a sword blade. Yes, there's a lot more craft skill detail involved, but the core process is that simple: the twisting makes the complex patterns.

The second point that was bugging me involved the pronouncements Voynich experts had made regarding the manuscript's low-level nuts and bolts issues. A lot of people were guesstimating numbers for how long the manuscript would take someone to produce, or how easy it would be to produce a particular feature. Nobody appeared to have sat down with pen and parchment and tried some field experiments. To me, several of the odd features of the text in the manuscript looked suspiciously like side-effects from some low-level practical feature of how the thing was produced. For instance, words toward the end of a line tend to be shorter than words near the start; similarly, some characters of the Voynichese alphabet are more likely to be found in some positions in the line than in others. That, to me, shrieked of an accidental feature of the inky-fingered craft skill production process.

Finally, an outstanding Brazilian mathematics professor named Jorge Stolfi and his team had shown that the text in the manuscript wasn't the product of randomly combining gibberish syllables in mix-and-match fashion. There were nonrandom regularities in the text, enough of them to knock that idea on the head. I was already pretty sure that the text hadn't been produced by someone simply making up words, since the statistical regularities were hard to reconcile with that explanation. There was something missing in that picture, but I didn't know what.

Then one day it hit me. Stolfi had done a brilliant job of testing whether the manuscript had been produced the way almost all modern scientists would have put it together, if they wanted to produce gibberish in large quantities. He checked whether the syllables in the Voynichese words had been randomly combined, and he showed that they hadn't. What he and everyone else had missed was that when the manuscript was produced,

nobody would be using random combinations, since the concept of randomness wasn't discovered until centuries later.

Instead, the hoaxer would be producing *quasi*-random combinations. It sounds like a nitpicky detail, but it's a huge difference. At least one researcher has become rich by spotting that
some lottery scratch cards use quasi-random rather than random
numbers and working out how to pick the winning cards. An
effective example is the difference between honest dice (random
in the strict sense—every face is equally likely to come up) and
loaded dice (quasi-random; one face will come up more often
than the others, but it won't come up every time).

When I put those three points together, they all suggested
one thing. Maybe the manuscript was a meaningless hoax after
all, produced by combining gibberish syllables in some low-tech
medieval way that had probably been forgotten for centuries. It
was an avenue nobody had seriously traveled before, and I could
explore it without needing any cryptography expertise. All I
needed was some paper, a calligraphy set, and patience. Plus the
ability to figure out how and why someone five centuries ago
might have tackled the problem of producing large quantities
of meaningless gibberish, preferably as swiftly and efficiently as
possible.

I could see one possible part of the solution immediately.
Dee and Kelley had used tables of cryptic symbols in their séances; tables of figures and numbers were well established by
that period. Anyone learning Latin would be familiar with the
idea of breaking up words into prefix, root, and suffix, just the
way that it's done in old-fashioned grammar, and just the way
Voynichese words are structured. That practice lends itself readily to producing tables where some columns contain prefixes,
others contain roots, and others suffixes. That was an obvious

place to start. I could even make a fair guess at how big the table would need to be: wide enough to produce about a dozen words in one go, because that's the typical number of words in a full-length line in the manuscript, and deep enough to produce about forty or fifty lines in one go, the typical size of a text-only page in the manuscript.

There also had to be some device, some mechanical gadget or quasi-randomizer to select the syllables for each word. True dice were out of the picture because they would have produced random distributions and been detected by Stolfi's work. Back in the premodern world, truly random dice were rare; it's difficult to produce dice that are completely fair, because of the differential weight that results from drilling more holes in some faces than others, and differential density if the dice were made of natural material like bone or wood, which are not homogeneous all the way through. The same went for methods inspired by other games of chance popular when the manuscript was produced, like flipping coins. I started reading up on old technologies, looking for insights.

And then my subconscious popped up an unexpected answer, by using the old trick of taking something and reversing it. The thing that my subconscious flipped around was a Cardan grille. The Cardan grille was invented in 1550 by Gerolamo Cardano, an Italian polymath. He was a character so colorful that he dominated any ordinary story, but in a setting as outlandish as the Voynich Manuscript story, he's no more than a bit player. His invention was deceptively simple: the grille was a sheet of stiff paper or metal with holes cut in it. The user laid the sheet over the writing paper, and wrote the intended message where the writing paper was visible through the gaps in the overlying sheet. The sheet was then removed, revealing the intended message surrounded by areas of empty paper. The user then wrote innocuous text to fit around the real message. To read the message, the recipient laid an identical copy of the grille over the page, and read the words revealed through the gaps in the grille. So, for instance, the intended message might be, "The enemies are planning war," written through two gaps in the grille; the space between them might be filled with text such as "in no way," producing an apparent message of "the enemies are in no way planning war." The Cardan grille was extremely popular from its invention until the end of the century. By that time, though, its limitations were becoming obvious. A principal limitation was that joining the innocuous text to the real message was difficult—the text was often slightly misaligned, and it frequently required tortuous phrasing to join innocuous text with the real message in a grammatical way that made reasonable sense.

Once you knew what to look for, the grille's presence could be detected, and so the method faded into oblivion. It enjoyed a brief resurrection during the American struggle for independence centuries later, when Sir Henry Clinton, a British general, sent messages back to London using this device: by then it had been so widely forgotten that American intelligence agents appear not to have suspected its presence in captured documents.

I had been thinking about the Cardan grille because of an odd feature on some pages of the Voynich Manuscript, which a brilliant Oxford graduate named Philip Neal had noticed. He had spotted that the alignment of the lines on those pages was strange, as if someone had written lines 1, 3, 5, and so on down the page, then gone back and filled in 2, 4, 6, and so on, subtly misaligning them with the first set. I had wondered whether that was an indication of the Cardan grille being used to encode real, meaningful text. I had eventually abandoned that idea, for various reasons, but it had stuck in my subconscious.

What if someone had used a Cardan grille the opposite way around—to produce meaningless gibberish instead of to encode meaningful text? It was blindingly obvious with hindsight, and very easy to test out. By the end of that evening, I'd found my answer: a method that could produce what everyone had thought was impossible: text with the same complex regularities and previously inexplicable odd features as the text in the manuscript. I had found the most reasonable explanation yet for a four- or five-hundred-year-old mystery on a budget of under thirty dollars, in a few weeks. And I had found it using Verifier, without realizing it at the time. We had our proof of concept.

7

Dissecting
Verifier

AFTER A FAIR BIT OF WORK, I HAD AN INTERESTING
answer about the creation and purpose of the
world's most mysterious book. This immediately
raised a lot of questions.

If I was going to share what I had discovered with the rest
of the world, I had to be able to answer these questions. The
answers had important implications for the rest of our new
problem-solving approach, Verifier. Simply put, I needed to
know that there was a solid scientific basis behind my analysis
of the Voynich literature. If I couldn't pick up the tools and
apply them to *other* unsolvable problems, Verifier wouldn't get
off the ground.

You've just read the equivalent of Voynich Lite—my adven-
ture with the manuscript without the Verifier details. Now it's
time to look more closely at how I applied the Verifier Method.
First I'll show you the tools I used—among them our old friends

drawing a picture and *gathering hands-on experience*—and then I'll unpack each of the Verifier chunks.

Tables and Grilles

Tables

One part of my solution came straight out of the core literature on the Voynich Manuscript. There was pretty solid agreement among researchers that Voynichese words can be divided into three parts: prefix, root, and suffix. That's an obvious place to start looking for ways to produce a meaningless hoax: you can simply draw up a list of prefixes, roots, and suffixes, then look for a way to combine them. But there were some strange features of Voynichese to take into account when working with the lists previous researchers had compiled.

One odd thing about Voynichese is that regular patterns govern which syllables occur where, but those regularities aren't quite the same as in most ordinary languages. For instance, in Voynichese, a word can consist of the full set: prefix + root + suffix. But it can also consist of other mixtures. Some words are just prefix + root, or root + suffix, which are structures you see all the time in ordinary languages—think of the words *undone* and *going* in English, which have those two respective structures. Here's an example of how you might try to generate meaningless English-style words using this approach (figure 22).

If you mix and match pairs and triplets of adjacent elements, you can put together a large number of words that could be described as "English-like." Some of the words *would* be real English, such as *unseeing* or *singer*. Others would be plausible-looking: "refeelable," for example. But Voynichese allows structures you'd never see in English, such as prefix + suffix.

One of the most popular pages of the Voynich Manuscript—folio f33v—depicts what appear to be two flowers, with leaves and strange, tuber-like roots. If the image is of sunflowers, that would date the book to sometime after 1492, since sunflowers are New World plants. For cryptographers, f33v raises further interesting questions: How did the book's author create this page? Which came first—text or flower? And do the answers assist us in deciphering the mysterious book?

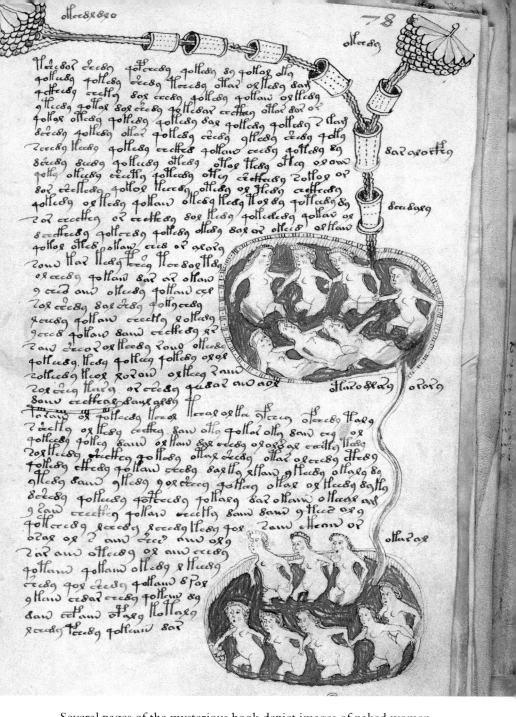

Several pages of the mysterious book depict images of naked women bathing in pools surrounded by strange plumbing. Nobody knows what the odd structures are meant to be, or the purpose of the bathing.

An original plant image from the VoynichManuscript (above) and the author's reproduction. Researchers often insist that the text and artwork would be too complex and take too long to be worth a hoaxer's effort. But this reproduction took less than two hours to create. Duplicating the entire book at this rate would take about ten weeks—a reasonable trade-off if a con man were expecting to sell the book for a profit to a wealthy mark.

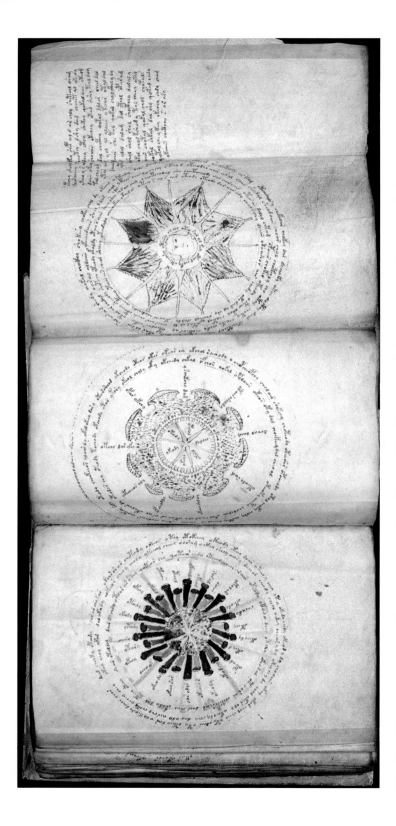

The Voynich Manuscript is problematic on a number of levels. This foldout—depicting astronomical or astrological images—shows how just numbering the pages is a challenge. Does this constitute one page, or four?

The author's proposed solution to the Voynich mystery uses a table (the large grid of syllables) and a grille (a piece of card with three holes cut out of it). By varying the placement of holes, and using a couple of different tables, one could easily create a sizable book of nonsense text. Ink blotches like the one in the upper left are a familiar by-product of the technology. The Voynich Manuscript's author may have worked around them, accounting for some of the text's regularities.

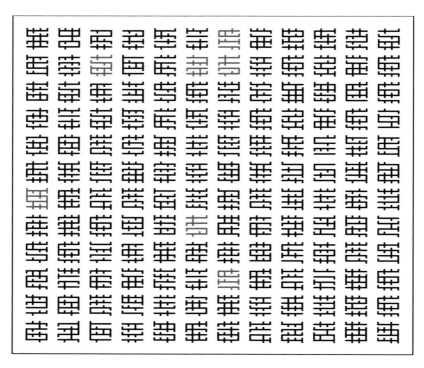

Inspired by the Voynich Manuscript, the author has created two still-uncracked codes. This page from the sixteen-page Penitentia Manuscript shows how codemakers can design text to frustrate code breakers. The design of this page, for instance, does not give you any clues about which direction you are intended to read—top to bottom, left to right, or some other way. The complete ciphertext is available at: www.hydeandrugg.com.

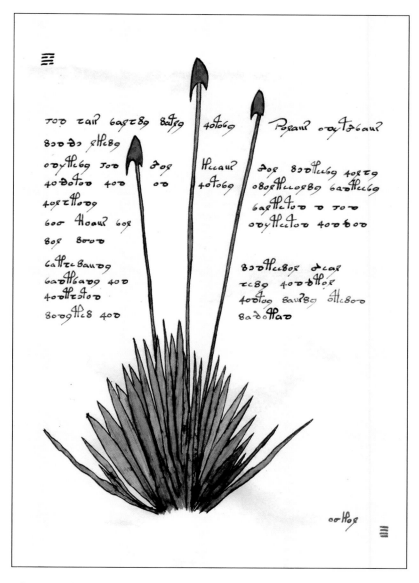

The Ricardus Manuscript, also created by the author and as yet uncracked, shows a completely different type of code: a Voynich-like
manuscript that does contain a coded message. The complete ciphertext
is available at: www.hydeandrugg.com.

The author's Search Visualizer software allows researchers to scan huge chunks of text, even if they don't speak a word of the language, and discern meaningful patterns. This illustration shows that Genesis, the first book of the Bible, has patterns in common with the Gilgamesh epic. Red squares show where the word "life" appears in the text; green squares show where the word "death" appears.

Prefix	Root	Suffix
un	see	er
dis	go	ing
re	feel	ed
de	sing	able

FIGURE 22

That's getting toward the weird side; the equivalent of *uning* in English, a combination you're not going to see unless the author has been drinking heavily before hitting the keyboard.

In the manuscript, Voynichese prefixes often occur on their own, the equivalent of *dis* on its own in English. Voynichese suffixes also frequently occur on their own—equivalent to *ing* as a freestanding word in English. It makes you wonder why anyone would create words with such a weird structure. One possible explanation is that the book's creator didn't know it was a weird structure, but that explanation is unlikely; if you know enough about word structure to produce this method, then you also know enough to spot how odd Voynichese structure is.

Another explanation is that the book's creator didn't care. This explanation is more plausible. The Voynich manuscript looks like something hacked together by somebody whose mind was elsewhere. As a point of comparison, try looking at any professional-quality handwritten document from the 1400s or 1500s. Those documents have neat, straight lines of text, carefully aligned with the margins. The Voynich Manuscript doesn't. It was written freehand, with little attention to margins. Most of the illustrations look as if they were done by a talentless apprentice, and the coloring-in is at about the level you'd get from a bored ten-year-old.

The usual assumption in the Voynich research community is that the manuscript was a working journal created by some medieval scholar, intended only for his or her own use, and therefore a scruffy document. That was a fit with its appearance. There were other possible explanations, though. One was that if you were a hoaxer hacking together a fake book to sell to a rich mark, you wouldn't want to spend any more time on it than you had to. The unique selling point of the manuscript in that scenario would be the incomprehensible script twinned with the bizarreness of the illustrations; adding artistic quality to the illustrations and calligraphy probably wouldn't add much to the sale price.

It is also possible that the manuscript's creator had never learned the skills of the professionals who produced the top-end manuscripts of the day. If you were a hoaxer faced with that problem, pitching the book as an undeciphered notebook containing the secrets of some unknown alchemist would take a problem and turn it into a strong point, adding plausibility.

Thus the odd syllable structure of Voynichese might be simply the result of a cheapskate hoaxer cutting everything to the bone. I wondered, though, whether there might be another explanation. If you look at a document in any real language, you find words of different lengths. That's so familiar that most people don't consciously notice it. If you're a hoaxer, though, and you're trying to produce something that looks reasonably similar to a real undiscovered language, it's a big potential problem.

At the time I was doing this work, the most likely suspect for a hoaxer (though by no means the only one) was Kelley, and one of his experiences with Dee raised some questions. Early in their acquaintance, Kelley produced an alleged treasure map containing coded instructions and showed it to Dee, claiming that it was a genuine map he had acquired. Dee cracked the map's code without much trouble. After that experience, Kelley would have known that Dee was not an easy mark. If Kelley did indeed produce the manuscript to sell to the gullible Emperor Rudolph II, then Kelley would have been well aware that Rudolph's next move would probably be to ask Dee to decipher it. Dee knew a lot about the comparative frequencies of letters in different languages, since that was (and still is) the standard way to begin cracking a code. So anyone trying to produce gibberish that looked like a real language would need a way of generating "words" that varied in length and contained a spread of common letters, medium-frequency letters, and rare letters.

If you look at the syllables of Voynichese from that viewpoint, you start seeing patterns that look like someone has been trying to do just that: produce different word lengths and different letter frequencies.

Here's an example of some common prefixes in Voynichese: *o or oro ol olo.* If you map them out one above the other, you start to see a possible pattern. It looks as if someone is putting together short, medium, and long syllables systematically.

o

or

oro

o

ol

olo

It's not completely systematic, but it's a strong trend. The same trend occurs in the suffixes. Here's what a section of a mix and match table for Voynichese 1.0 might look like (figure 23).

That's version 1.0. As we saw earlier, though, a lot of Voynichese words don't have all three components; some have prefix + root, while others have prefix + suffix or just root. If you want to produce that pattern, you can do it by leaving some blank cells in your table, like this (figure 24).

So, with version 1.1, you can mix and match a blank from the prefix column with a *che* from the "root" column and a *dy* from the "suffix" column, and you get the Voynichese word *chedy,* which is fairly common in the manuscript. It's a promising start. But once you've worked out how to create words with

Prefix	Root	Suffix
o	che	d
ol	chee	dy
olo	she	l
or	shee	ldy

FIGURE 23

Prefix	Root	Suffix
o	che	d
	chee	dy
olo		l
or	shee	

FIGURE 24

this approach, you run into a new problem. In Voynichese, as in all real languages, some syllables are very common, like *ing* in English, and others are moderately common, while some are rare. If you're planning to replicate this spread of syllable frequencies in Voynichese, where some syllables only occur once every few hundred words, you need a way of keeping the frequency of those syllables consistent for page after page. One easy way to do that is to think big and create a table large enough to hold hundreds of syllables.

Here's what a small section of such a table could look like (figure 25). Pick from the first three columns to produce your first word, then pick from the second set of three columns to produce your second word, and so on. A dozen sets of three columns would let you produce a typical-length line from the Voynich Manuscript if you moved your grille across the table, generating one word at a time.

The table I produced had several hundred cells. I did it on replica parchment, with pen and ink and ruler. If you want to tap into the mind of a medieval hoaxer, you have to do things

Word 1			Word 2		
Prefix	**Root**	**Suffix**	**Prefix**	**Root**	**Suffix**
o	che	d			dy
	chee	dy	40	te	dol
olo		l	o	kee	
or	shee		sor	shee	y

FIGURE 25

the old-fashioned way. And while you do it, you have to listen to what the experience is telling you.

Hands-On Experience

Remember what we said in chapter 2: if you observe experts at work, you spot things you would not be able to easily elicit from them. The researchers who watched the seamen loading the bulk carrier spotted much that the ship captain had omitted from his interviews.

I could not watch the long-dead creator of the Voynich Manuscript at work, but when you create a big table of gibberish syllables using pen and ink, replica parchment, and ruler, and you pay attention to the low-level issues you encounter, you start noticing all sorts of unexpected things.

One thing that surprised me was how quickly you can produce a big handmade table. It's a boring process, particularly drawing the guidelines that mark out the rows and columns. That part took longer than I had anticipated. And you have to be careful not to smear the guidelines if you're working with a traditional ink pen rather than a modern one. But you can draw and fill in a table that size, enough for a dozen words across and forty lines deep, in just one morning. That's fast enough to make a gibberish hoax look like a viable business proposition, if you're thinking about return on investment.

The smears from using a traditional calligraphy pen that you dip into ink, rather than a neat, modern, low-smear pen, also threw up some unexpected insights. I had wondered why someone might think of using empty cells in the table as a way of changing word length and word structure. Within a few minutes of putting pen and ruler and replica parchment into action, you find one possible reason. It's horribly easy to smear the ink and obliterate a square. That's going to give you the idea of having empty squares, even if it's only because you can't face the hassle of letting the ink dry and then scraping it off the parchment the way ancient scribes did when they made a mistake.

I also started seeing regularities in what would almost inevitably go wrong when you tried to fill in the table, even if you were making a deliberate effort to fill it in so that there weren't any telltale oddities in its structure. For instance, suppose that you've decided your table should contain, say, fifty-five instances of the syllable *ldy*. How many does that work out to per column, if you have a table with fourteen columns set aside just for prefixes? Even if you're feeling fresh, you'd probably have trouble working that out and keeping track of whether you had reached fifty-three or fifty-four when you were filling it in, since the answer isn't a neat exact number per column. I found that the easiest way to fill those columns with syllables was one syllable at a time—for instance, doing all the *ldy* syllables in one go—and to fill in one column at a time. It's still not easy. You're trying to keep track of how many times you've used that syllable, and you can't simply put it into every sixth cell or whatever, partly because that would produce regularities that a code breaker would spot, and partly because when the cells start filling up, it becomes a challenge to find blank squares for your syllable. That had implications for any gibberish you produced using the resulting table. One implication is that words produced using the last few columns, where you would be most likely to have problems fitting in your syllables consistently, would tend to be different in length from words produced using the other columns. If you were working left-to-right across the columns with the grille to produce a line of text, that would mean words toward the start of the line would tend to be a different length from words toward the end. This is exactly what happens with the text in the Voynich Manuscript.

It's also easy to lose track of where you are on the list of syllables to put into your table. You get called away to handle some minor domestic issue like feeding the cats, and when you come back, you've forgotten how many times you've written the syllable *oly* in your table. Or you miss one syllable completely, and only notice the mistake when it's too late and there's only space for putting it in the table a handful of times, instead of it being a common syllable as you had originally planned.

None of this is a big deal if you're only using one table. But there's a limit to how often you can pick from the same table before obvious regularities start showing up. After a while, you need to create a new table, with the same syllable frequencies but with the syllables in different places (so you don't end up with give-away regularities in where they occur on a page). If you make any of those "losing track" mistakes when you do your second table, there's a real risk that the second table will have a different frequency of some syllable that's common in the first table, resulting in text that looks like it's in a different dialect. Which is what you get in the Voynich Manuscript.

You also get interesting effects from rare characters. Say you have a syllable containing a rare character, which only occurs once per table, near the end of your table. Suppose again that you're producing each line of text by working across the columns in your table. That rare character tends to happen near the end of the line on a finished page. Which is exactly what happens with some rare characters in the Voynich Manuscript.

There's another side effect from creating words in this way. If you're using a fairly systematic way of creating the prefix, root, and suffix syllables, the words formed by combining those syllables are likely to vary systematically in length, with a few words that are very short, a few that are very long, and most of medium length. There's a good chance that this frequency distribution will take a form known technically as a binomial distribution. That distribution of word lengths is very rare in real languages. However, it occurs in Voynichese.

All these features had been noticed by previous researchers. But nobody had a good explanation for them. If the manuscript had been produced using tables of syllables, though, those features were not only easy to explain; they were almost inevitable. They were so predictable that any hoaxer would realize that they had a new problem to solve. To hide an effect like this, which would be a lot more obvious to a critical mind back then than to a modern one, you would need one other tool.

Grilles

If you're a modern researcher, you're taught to work systematically. That fact is taken so much for granted that it's often not mentioned explicitly; it just never occurs to you that anyone could work any other way. It's a particular issue in cryptography, the field of code making and breaking. If you don't encode your message systematically, the person at the other end won't be able to decode it and understand it.

There's a strong suspicion that this is why the D'Agapeyeff Cipher has never been cracked. Alexander D'Agapeyeff wasn't an expert cryptographer by background; he wrote a textbook on the subject because his publishers wanted a book about it, and they saw him as a dependable author in his main field, cartography—mapmaking—with enough knowledge of cryptography to write a basic introductory textbook. So it's plausible that when he produced the now-famous cipher text that bears his name, he made a mistake partway through, leaving the text indecipherable.

So if a modern researcher was looking for a way to mix and match syllables using a table, the automatic reaction would be to look for a way of doing it systematically. The first thing she'd do is reach for the random number generators. That's not how it would have been done when the manuscript was created, however, because the concept of randomness hadn't been invented when the manuscript was produced.

One obvious solution—with hindsight, at least—is to use a tool that shows you only three syllables of the table at a time. One such tool is a version of the Cardan grille, the encoding device created by Cardano, whom we met in chapter 6. You take a card with three holes cut in it, each the same size as a cell on your table, and lay it over your table. The three syllables it reveals give you a word to copy onto your own document. By varying the patterns of the holes cut into your card—your *grille*—you can reuse the same table over and over again, generating new text each time.

Here's what a couple of examples might look like (figures 26 and 27). The gray cells show where holes would be cut through the card. (See the real ones I used in color figure 5b.) You'd

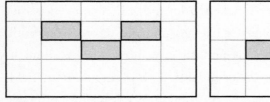

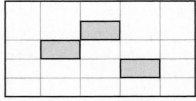

FIGURE 26   **FIGURE 27**

need a lot of these cards, each with a different pattern, to vary the word selection from the same table.

Here's what you get if you put the first grille over the table we produced earlier (figure 28). It shows the sets of three cells with text in bold. I've grayed out the cells that are hidden by the paper of the grille. This grille gives you two words: *ocheed* and *tedy* (there's an empty cell at the start of *tedy*).

The second grille gives you this different pattern from the identical section of table (figure 29). Remember that the gray

	Word 1			Word 2	
Prefix	**Root**	**Suffix**	**Prefix**	**Root**	**Suffix**
o	*che*	d			dy
	chee	*dy*	*40*	te	*dol*
olo		*l*	*o*	*kee*	
or	shee		*sor*	shee	*y*

FIGURE 28

	Word 1			Word 2	
Prefix	**Root**	**Suffix**	**Prefix**	**Root**	**Suffix**
o	*che*	d			*dy*
	chee	*dy*	*40*	te	*dol*
olo		*l*	o	*kee*	
or	shee		*sor*	shee	y

FIGURE 29

cells are masked by my grille; the white cells are the only ones the hoaxer would be able to see.

You now get the words *olochee* and *otey*. Same table, different grilles and words. But there's a potential problem, if you're expecting your text to be scrutinized by someone who knows about codes. If you just move this piece of paper—the *grille*—systematically from left to right across the table, three cells at a time, you start to get noticeable regularities. For instance, you might find that a particular rare character only occurs in the third word on a line or in the fifth line down on a page.

When I made my discovery public, Voynich researchers were interested in the idea up to this point. Their next question, though, made me do a double-take. They wanted to know the exact algorithm my alleged hoaxer had used to hide the regularities. Being used to working systematically, they were expecting an answer like "Move the grille along horizontally for four words, then down one row for the next five words, then up two rows for the remaining words in the line." But what I was saying was the exact opposite. If you're trying to conceal regularities in how your hoax was produced, the last thing you want to do is to use a regular, systematic method—an algorithm—for introducing those irregularities. If you do, there is an excellent chance that a code breaker would spot the algorithm and unmask the hoax.

I never managed to get my point across. All the way through, I kept encountering that fundamental misunderstanding. The Voynich experts were taking it for granted that the method should be applied the way a cryptographer would apply a method for creating cipher text: apply a rule so it can be replicated by someone else. They just couldn't accept the point that a hoaxer would be thinking just the opposite: use no rule, so *the text can never be replicated by anyone else*. After all, that's the entire point of a hoax. Make it real-looking but impossible to crack, to heighten the mark's curiosity.

They're two very different paradigms.

Publication

There was, and still is, a well-established community of researchers studying the Voynich Manuscript. Some of their websites, such as Jorge Stolfi's and René Zandbergen's, are invaluable resources for anyone studying the manuscript. They have an honorable tradition of sharing their findings with each other online.

I had a different endgame in mind. Their goal was to unlock the secrets of the book. I wanted to apply Verifier to big problems—Alzheimer's, dyslexia, autism, malaria, and so on. It was a long, long list. If I was to convince anyone to fund Verifier research into one of those problems, I needed some peer-reviewed evidence that I was able to solve hard problems, and the Voynich Manuscript was my best shot at that.

Not many papers with proposed solutions for the Voynich Manuscript had made it through peer review since the 1920s. There have been plenty of books, essays, and blog articles, and several descriptions of the manuscript that appeared in journals. But only three articles proposing solutions had made it past expert reviewers and been published in reputable journals: William R. Newbold's paper in the 1920s, Leonell C. Strong's in the 1940s, and Robert S. Brumbaugh's in the 1970s. Each solution was swiftly demolished by experts in the field, and after that, journals became wary of the subject.

But for this to be a credible case study that I could talk about in other disciplines, I needed to have a peer-reviewed publication in a journal, preferably a good one. I went for *Cryptologia*, the leading journal on historical cryptography. The process of submission and review ended with the hoped-for verdict: acceptance, subject to minor changes.

I knew the story was likely to get attention, but it was anyone's guess whether it would get limited local coverage or something bigger. My university's press officer, the unflappable Chris Stone, helped me produce a press release, timed to go out just before the relevant issue of *Cryptologia* was published.

The story went viral. There were features in magazines and newspapers around the world. With hindsight, that's hardly surprising: it's a story that has everything. There's mystery and treasure maps and codes and Dee and Kelley and modern-day code breakers and all the rest.

I argued that the simplest explanation was that the manuscript didn't contain a code after all. If the possible explanations are an incredibly strange language, an incredible code, or a meaningless hoax that can be produced efficiently using ancient low-tech methods, then the most plausible explanation is the hoax. That implied they had all been looking in the wrong place for years.

I had been thinking of the Voynich Manuscript puzzle as just a puzzle that I had managed to solve using tools I knew. It had been a swift process, and I had been too deeply focused on it to see what I was actually doing. It was only afterward that I started to realize that, without realizing it, I had tackled the Voynich problem using Verifier.

The Verifier Connection

When you shift gears while driving, you aren't aware of which muscles you're using; you just change into the new gear and drive on without further thought. When you recognize a friend at the office, you don't know which bits of neural circuitry are telling you that this is Jane from my department, not Mandy from Accounts, who is much the same age and height and hair color. You do it, but you scarcely notice you're doing it, and you'd be hard-pressed to explain just what you're doing or how you're doing it. I had been doing the same with Verifier after working with the component parts of the approach for years.

When I tackled the Voynich Manuscript, I was using a lot of highly practiced skills from my own expertise in human-error analysis. Almost all the individual tools and methods I was using were also used by other researchers. But what I hadn't noticed was that I was using them in combinations that hardly anyone else would use: Verifier combinations. And I hadn't noticed it because of our old, easily overlooked friend: taken for granted knowledge, the unthinking assumption that everyone else knew what I knew.

The first five steps I'll enumerate for you, for example, used a lot of swift pattern matching, where I was simply scanning diagrams or batches of text, trying to spot the underlying structure, or message: *What is this author or research getting at?*

Here are the Verifier combinations that I used.

1. Analyze the expertise brought to bear on the problem. As I read through the articles written by other researchers, one of the things I was looking for was what types of experts were involved in tackling the Voynich Manuscript. If you look at the introductory sections to journal articles on a given topic, you hardly ever see a discussion of whether the previous research included work by experts in all the relevant fields. Instead, you'll see a historical account of which main players tackled this particular topic in the past and what the key findings are. You won't see them ticking off the list of relevant fields of expertise and thinking about which fields are missing. That's a different way of thinking about the problem.

When you perform an expertise check on the Voynich literature, it quickly becomes clear that the "unknown language" strand and the "unbroken code" strand had been investigated by high-quality experts in the relevant specialties. As for passing judgment on the "complexity" argument, people like Stolfi, an expert in engineering, mathematics, and computer science, would know a lot about some aspects of complexity theory. But they wouldn't know about the specific area of complex outputs from low-tech equipment.

2. Determine which groups of people are studying the problem, what methods they use to tackle problems, and what forms of representation and analysis they use on the knowledge they collect. It's staggering how differently research is done across and within disciplines, even when the researchers are working on closely related areas, or on different aspects of the same area. There have been a lot of studies of "communities of practice"—people who work on the same area—and "communities of discourse"—people who talk to each other about the same area. The Voynich community is a prime example of how those differences play out.

The main Voynich community was located in the blogosphere and ranged from teenage hobbyists to legendary cryptographers. Some of this community's work describing the manuscript had been published in peer-reviewed literature, but most of it was published in the online mailing list. I noticed that there was very little publication of empirical, data-driven work: very few people were quoting reliable, hard figures for, say, how long it would take a competent professional medieval scribe to write five hundred words of legal text as opposed to high-end illustrated manuscript.

In the realms of visual representation and analysis, there was a wide array of expertise; for instance, Phillip Neal is an expert on medieval documents, and Jorge Stolfi is a professor of mathematics. The key insight Stolfi focused on—whether the text was *random* gibberish—was well within his area of expertise. But there were also expertise gaps.

3. Collect data from the published literature. Jo Hyde and I had assumed that this strand would be huge. But in fact, I didn't find this stage of the Voynich project as grim as I had feared. Part of that was luck; the threads led into fields where I had at least the equivalent of a phrasebook-level grasp of the relevant concepts. Part of it was because, when you're at this stage of tracing threads, you're not yet systematically testing the strength of the threads. Thus if you see one reasonably sane paper asserting a particular statement, you don't need to wade through every other paper that makes the same claim, because that would be needless duplication of effort. So I tracked the Voynich threads into the literature on talking in tongues and Enochian and Martian and a whole batch of other places, and they all dead-ended quickly. There wasn't a code that produced texts like Voynichese, there wasn't a language with the features of Voynichese, and there wasn't some other form of human output, like channeling or speaking in tongues, that produced anything quite like Voynichese.

4. Collect knowledge from people to fill gaps in the literature. Theoretically, at this stage a Verifier team would now fan out to unpack knowledge from various experts in the field. I didn't need to do this in the Voynich case study because there weren't any gaps that I could fill using that approach. The obvious gap was that no one with the proper expertise had investigated the hoax hypothesis, and it was clear that there were no human experts I could quiz about how to do a medieval-style hoax that would look like a mysterious cipher text.

5. Use your knowledge of how experts operate and of human error to spot key assumptions and reasoning that might be in error. Jo and I had visions of spending years painfully annotating papers and interview transcripts, then going to the experts and bugging them to distraction with questions like "What do you mean by this phrase here, in this context?"

Instead, in the Voynich investigation, this stage threw up a key assumption without any need for annotation or transcripts or interviews. The "too complex for a hoax" assumption jumped into center frame as soon as I started looking through the previous work with my Verifier hat on. This was a classic weak link in the chain of reasoning. It's not enough to say something is too complex for a hoax. You must have evidence to support such an assertion.

6. Choose the appropriate formal logic for each key assumption you've spotted, and apply each form to each assumption, checking rigorously for errors. I didn't get to that stage with the Voynich case study because I had already identified a weak link that was a top priority for more investigation. Testing that link meant that I would need to get my fingers inky, using ancient technologies to see whether I could produce something like Voynichese as a proof of concept.

Within a few weeks of starting to use those technologies, I had found a way of producing something that was a lot like Voynichese and could be used to produce a book the size of the Voynich Manuscript in a few weeks—that is, fast enough for a hoax to be easily feasible. In the end, I had produced a significant insight into a field where I had no previous expertise, using Verifier.

One of the reasons Verifier has such huge potential is that it draws heavily on well-established tools from a wide range of fields; another is that it has those tools organized into the conceptual equivalent of drawers, with a set of guidelines on each drawer about *which* tool to use in *which* situation. If you don't ordinarily use a particular tool in your line of work, you probably don't even know that it exists. And that can hamper the solution of difficult problems. Verifier makes those tools explicit for experts, researchers, and decision makers, packing them into the method's third main feature: the toolbox is big and heavy.

Research methods are usually difficult to explain in abstract terms and can sound like Zen abstractions, so we use a lot of concrete examples to illustrate the tools in the Verifier box. One of the most striking examples is a long-established favorite of experienced researchers: *taking the obvious and reversing it.* Here's an example of how military researchers used that method to produce a better torpedo via a design that initially appeared to make no sense at all.

All navies want fast torpedoes, and for years nobody had them. Naval designers tried the obvious—improving the streamlining, tweaking the propeller, installing more powerful engines. Whatever they tried, they kept hitting the same physical constraints about how fast a torpedo-size hunk of metal could move through water: about fifty knots (nautical miles per hour). To put that in context, the current speed record for a wind-powered sailboat is just over fifty knots. NATO military engineers during the Cold War consoled themselves with the knowledge that their Soviet counterparts would be constrained by the same laws of physics, so the Soviet technology couldn't be radically better.

And then perestroika happened and the Berlin Wall came down, and the NATO experts discovered that their Soviet counterparts had taken the obvious and reversed it. They'd asked, in effect, "What happens if you do the *opposite* of making the torpedo streamlined, and then give it a lot of power?" To most people, that sounds like a pointless question—unless you pose it to a marine engineer, whose response after a moment's stunned silence would be, "You'd get cavitation."

That's what happens when you're trying to push an object through water, but you're doing it so fast that the water can't move out of the way in time, so you get bubbles. Cavitation bubbles are a nightmare for propeller engineers, since they abrade the surface of the propeller and also make telltale noises that hostile navies can use to identify not only the type of craft you're in but often the specific individual sub or ship.

In the context of torpedo design, however, the cavitation bubble phenomenon changes the game completely, as the Soviet engineers realized. They understood that if you get enough cavitation at the front of the torpedo, the torpedo will be flying through a bubble, not swimming through much denser water.

How do you get enough cavitation? Easy. Stick a rocket motor on the back of the torpedo and build a torpedo nose that's about as subtle as a brick. The nose plows into the oncoming water, throwing up a bubble that surrounds the torpedo, and the torpedo shoots through the bubble and the surrounding water, exactly like a missile through air.

Does that actually give you more speed than the fifty-something knots of the NATO torpedoes? You could say that.

Officially, the standard Soviet torpedo based on this technology, the Shkval, does noticeably better than fifty knots—in fact, the official speed is two hundred knots. Unofficially, there's good evidence that it can do two hundred and fifty knots, and there are rumors of a version capable of *three hundred* knots.

It's probably just as well that Hannibal never got to meet that Soviet design team, or world history could have taken a very different direction. . . .

Aftermath

I had shown that the manuscript could be a hoax, containing nothing but meaningless gibberish. That's not the same as showing that it *was* a hoax or showing that it contained only meaningless gibberish. I was trying to get used to the idea that I would probably never know the answer. It was a frustrating prospect. And then, out of the blue, several years later, when we were deep in a second Verifier case study, came a couple of tantalizing new pieces of evidence.

An Austrian nuclear physicist named Andreas Schinner found a way of working out whether the text in the manuscript was more like a hoax or more like meaningful language. It involved some heavy statistical concepts that were normally used in nuclear physics. Fortunately for nonstatisticians, the title and abstract of his paper, published in *Cryptologia* in 2007, spelled out his conclusion clearly. Schinner's results suggested that the text consisted of meaningless gibberish, not an unidentified language or cipher text. It wasn't a definitive answer, but it showed a way to finding one and gave a strong indication of what that answer would be. Suddenly, the Voynich Manuscript was back in my life.

I didn't particularly want to have the Voynich Manuscript back. There were a lot of more interesting things going on, including our second Verifier study. One of the things that surprised me most about the Voynich case study was that I had found a possible solution before even starting on the formal stage. I had assumed that it was just a huge stroke of luck. Now we were starting to see the same pattern in the second study, and we had to do a serious rethink. That's what I wanted to focus on. Instead, I was having to deal with the manuscript again, and it was throwing up new questions as well as answers to old ones.

One answer provided support for the hoax explanation from a direction that nobody had anticipated. An Austrian film company that was doing a documentary about the manuscript wanted to interview me about my work, on location in Europe. Somehow the Austrian team managed to get permission to run tests on the manuscript. One set of tests gave a date for the vellum on which the manuscript was written. That raised a complicated set of questions, which I'll cover in chapter 10.

A second set of tests gave a much simpler answer to *another* key question, which sprang from a conversation between myself and the Voynich researcher René Zandbergen, who was the presenter of the documentary. I met René during the filming and grew to like and respect him. While we were looking at an enormous projection of one of the pages from the manuscript on the film studio wall, we both had the same realization. The quality of the photos taken by the Austrian team, under Andreas Sulzer, was so high that it would be easy to see whether any of the pages in those photos bore signs of correction. That had huge implications.

If you're a scribe writing on parchment and you make a mistake, there's a standard way to correct it that has remained unchanged for millennia. You wait till the ink is dry, then scrape off your error with a sharp blade. Then you rub the parchment smooth and finally write the correct text on the smoothed area.

If you know you're producing meaningless gibberish, you won't need to bother with corrections, because you know that the text you're churning out is meaningless anyway. It's conceivable that an obsessive hoaxer might want to "correct" a mistake in the meaningless gibberish they are producing or have a reason for wanting a particular word in a particular place, but if there were no corrections, that would be strong presumptive evidence that the manuscript contained nothing meaningful

Therefore, knowing whether or not there were erasures would give us a key piece of evidence about whether the manuscript was just a meaningless hoax. The Austrian team was well aware of the significance of this issue. A few weeks later, I received an email from René telling me what they had found.

René's email said that that the team hadn't found any visible erasures and corrections, which is unusual for a document as informal as the manuscript; even fairly upmarket medieval documents often contain words visibly crossed out and corrected, where the scribe couldn't be bothered to scrape the parchment clean and rewrite. The Austrians knew what they were looking for, and they had beautifully detailed photographs of the manuscript. They had looked at a number of pages, finding no evidence of erasures on any of them. The implication was clear: either someone had written those pages without making a single mistake, or whoever had written them didn't care about mistakes. The first option was hard to imagine; even the scribes who produce handwritten Torahs, arguably the most meticulous scribes on the planet, sometimes make mistakes, which they correct by erasing them and writing the correct text on the scraped-clean section. The "didn't care" option doesn't make sense if the manuscript contained code, where the text needs to be error-free or it will make the code unreadable. It does, however, make a lot of sense if the manuscript contains only meaningless gibberish.

Either way, the Voynich Manuscript was in some ways a brilliant demonstration of concept for Verifier. It did have some shortcomings, though. There was the frustration of not knowing, and probably never being able to know, whether the manuscript actually *was* a hoax, as opposed to demonstrating that a hoax was feasible. It was also a case study completely unrelated to any real-world problem. Beyond shedding light on the Voynich Manuscript, the results couldn't be used to do something useful in another field.

Our course was clear. We had to prove that this wasn't just a fluke by tackling a second problem that had defeated experts for years. This time we tackled a proper medical problem, which turned out to be a brutal case study, making me realize how lightly I'd escaped with the workload for the Voynich study. My colleague Sue Gerrard did the vast majority of the work on that case, while I worked on a different aspect of error. I was looking at the way our aesthetic preferences subtly steer us to think in particular ways, even if those ways aren't actually correct. Many of those preferences relate closely to mathematical concepts such as ratios and proportions, so this area could be described as the mathematics of desire.

The Mathematics
of Desire

U
P TO NOW, THE HUMAN ERRORS WE'VE LOOKED AT involved people getting something wrong because they weren't able to work out the right answer. That topic has received a lot of attention for centuries from logicians and mathematicians and others. There has been far less work on people getting it wrong because of the converse problem: our brains are wired to make us actively favor particular types of answers, regardless of whether those answers are logically right or wrong.

If you're a normal human being, hearing that the Voynich manuscript is a hoax just doesn't sit right with you. Your gut tells you that there must be something more to it. Surely four or five hundred years of mystery can't end in such an anticlimactic way—can it? That unsettled, dissatisfied feeling you're experiencing is at the heart of this chapter. What happens when researchers don't realize that they're subconsciously searching not for the truth but for an answer that *satisfies* them? It's worth looking into this a bit before we return to the Voynich story.

In a nutshell, I'm very interested in the low-level inbuilt preferences that nudge us toward favoring some explanations over others, some questions over others, some ways of tackling problems over others. Such behavior—if it is indeed hardwired into the human mind—has far-reaching implications for the types of error we're predisposed to make. This chapter explores some surprising findings in this area of science, then closes by discussing how inbuilt human preferences might impact research and business—and ways we could even harness them to our benefit.

Some of the earliest work in this area was done by the ancient Greeks. They found some tantalizing regularities, but their explanations for what might cause those regularities were shaky at best. The most famous of their discoveries is the Golden Ratio, which is the number 1.61803399. It's special partly because of its mathematical proportions. If you draw a rectangle using arcs of circles, all sorts of classic ratios and numbers crop up, including pi, which is why the early geometers got excited about the Golden Ratio.

When the ancient Greeks first proposed that human beings were disposed to like the shape of this golden rectangle, whose proportions were based on the Golden Ratio, that realization meshed well with a whole way of thinking about numbers and aesthetics. The ancient Greeks were aware that harmonies in music could be described mathematically, and that harmonies were also about proportions and ratios. A lot of other proportions and ratios intersect with people's aesthetics, once you start looking. The Parthenon and the other Greek temples are full of them. If you want to see ancient evidence of this, there's a long-ruined temple where you can still see cut into the floor the mason's marks, which they used as a giant draftsman's board to mark out the full-sized outline of each piece of stone.

The tradition of deliberate ratio and proportion in design stayed alive down through the Dark Ages and is still alive today. Chess pieces are a good example. If you look at the formal design for one of the most famous chess sets, it's full of deliberate use of ratios within each piece—the breadth of the queen's crown compared to its height, and the height of the crown compared to the height of the whole piece, for instance. That's just one example of proportions in design; there are thousands more, from the typeface in this book to the humble building brick.

For over two millennia, the received wisdom has been that the Golden Ratio provides the most aesthetically pleasing set of proportions. Passed along for thousands of years, that wisdom is accepted by many today as an inarguable fact. But the ancient Greeks weren't experts in systematically gathering information about people's perceptions. When twentieth-century experts in that field investigated the question, they found that, yes, humans did like that ratio on the whole, but there wasn't a dramatic difference between their attraction to *that* ratio and their attraction to *other* ratios near it. People clearly do have aesthetic preferences, so if we can factor those into buildings and into the designed world in general, that should make the world a more pleasant place. In her book *Rats and Gargoyles,* the sci-fi/fantasy author Mary Gentle details a closely related issue, the interaction between requirements and design—principles like designing the steps for a medieval-style market square so that they're at a comfortable height for tired shoppers to sit on and including colonnades on a building exterior for shelter from the summer sun. Every shape and size and element of a structure has its purpose, intended to fit seamlessly with the needs and activities and desires of the humans for whom it is built. That carries right down to the size of the individual bricks: the traditional brick has a width chosen because it fits comfortably within the spread of the bricklayer's hand.

Proportions have been a key part of architecture and art through the centuries, so much so that the paintings of one famous artist ended up featuring in one significant early achievement of artificial intelligence (AI). The Dutch artist Piet Mondrian produced a range of striking canvases featuring geometric shapes with elegant proportions and colors. Because those shapes were easy to model with the software of the time, a team of AI researchers decided to see if they could simulate Mondrian's art by using software to identify the underlying mathematical regularities in his designs—his preferred sizes, ratios, and colors. The result? They found that a significant proportion of people actually preferred the Mondrian simulations to the real thing.

The implications were far reaching. The image rights for a major piece of art are worth a fortune. If software could produce art that people preferred to art by a great master, that would be a serious business proposition. There were other implications as well, for design and for evaluation of products. If software could figure out what people would like aesthetically, you could use that to assess or predict human response to a product before actually building that product, thereby avoiding the risks involved in exposing that prototype to outside view via market research. For anyone worried about a new design being leaked prematurely during market research or being stolen by a competitor, this would be an attractive option.

More ambitiously, perhaps could you use a combination of two pieces of software, one of them tweaking designs semi-randomly, thousands of them per minute, while the other assessed each candidate until between them they found the new killer design. That's just about feasible with present-day technology, if you know how. Again, it's a very tempting possibility, and one that the aerospace industry is already using to find the best designs for aircraft fuselages and turbine blades.

As for why software might be able to produce art that people prefer to originals by a great painter, and what the story is behind the Golden Ratio and other mathematical regularities underlying human desire, the first part of the answer takes us to a different aspect of proportion and another favorite subject of artists through the years: human faces.

Portraits, Prototypes, and Archetypes

Artists know a lot about the proportions of the human face. If you're taking art classes, it's one of the standard lessons you learn. You'll probably be taught about the proportions of the ideal face at the same time as you learn about the proportions of the average face. If you learn to draw in a tradition other than traditional Western European, such as Japanese manga, the proportions you learn for the face, and for the body, will be different. There's one thing they probably won't bother to tell you about the ideal face, though, because they'll take it for granted—our old friend, taken-for-granted knowledge, once again. Idealized faces, faces that people prefer, are symmetrical.

When psychologists started systematically testing people's preferences in human faces, this is one example where unspoken artistic wisdom was borne out by the evidence. People do indeed prefer symmetrical faces to nonsymmetrical ones, and that holds true across cultures; it's not just a Western preference. Humans also like unremarkable, average faces. If you superimpose and average out a batch of photos, people usually prefer the composite photo to any of the photos that went into that composite. One possible explanation is that it reduces mental processing, compared to the work your brain must perform to view and comprehend an unusual face.

The preferred explanation in antiquity was that this preference sprang from some deep, mystical Platonic ideal born in the human soul, but that theory was undercut by the scientific finding that *insects* prefer symmetry, too. It's starting to look like preferences for symmetry go deep into the animal kingdom, and hence aren't a product of human spirituality or human cultural convention.

Symmetry is usually a good biological signal that an animal is healthy, and therefore a good potential mate. It's probably no accident that in animals that have significant asymmetry in their bodily organs, such as us humans with our hearts skewed to one side and our stomachs off-center, the asymmetrical parts are kept well hidden under the skin. Visible organs, such as eyes, ears, and noses, on the other hand, are carefully symmetrical. Asymmetry is often a sign of ill health, such as a tumor or injury, and tends to be shunned—that's what made Richard III such an easy target for Tudor propagandists, with his asymmetrical shoulder.

But though symmetry may make for pleasing design, it doesn't always make for *efficient* design. Experts, like everyone else, had generally taken it for granted that aircraft had to be symmetrical. Burt Rutan is an expert in aircraft design—he started designing them when he was a teenager—who realized that this received wisdom wasn't always true. There's a subtle but important difference between symmetry and balance. In symmetry the item is identical on both sides of the middle line.

In balance, the item has the same amount of weight or surface area, or whatever the key variable is, on both sides of the middle line, but it isn't identical in appearance on both sides. Sometimes, in aircraft design, what matters is balance, not symmetry. And sometimes asymmetry lets you do things you couldn't do with a symmetrical design. So in the mid-1990s, Rutan designed an asymmetric aircraft, the Model 202 Boomerang, designed to be safer in the event that one of the engines failed. The Boomerangs have a devoted fan following.

Rutan wasn't the only person to spot this gap in the received wisdom. In fact, humans weren't the only ones to come up with this approach. When propeller designers were trying to break out of conventional thinking and push propeller design significantly forward, some of them tried using artificial intelligence software known as "genetic algorithms." This type of software starts off with a spread of randomly generated solutions, assesses them, throws away the worst, tweaks the best, assesses these new versions, throws away the worst, tweaks the best, and keeps on repeating the process until it hits diminishing returns. One result that came out of this application was asymmetric propellers. Most human designers would never have dreamed of this solution, but those propellers were demonstrably better than previous designs. That often happens when you use genetic algorithms: you recognize that they've produced a great solution when you see it, but you realize that you would never, ever have come up with that idea.

Some artists deliberately break the rules of symmetry, intuiting that this approach makes for deliberately disturbing creations. The Swiss visionary H. R. Giger did the set designs for the movie *Alien*. His alien ship unsettles viewers because on the one hand it has a level of intricacy and detail and beaten-up realism that make it look very plausible, but on the other hand, it violates just about every comfortable assumption we have about what objects, especially spacecraft, should look like. It's a bizarre mixture of mechanical and organic; its interior spaces don't fit with any of our preconceptions of normal design; and it's *asymmetric,* deliberately very different from what we expect in a designed spacecraft. It's a world apart from the humans' ship in the movie, the *Nostromo,* which looks from the outside much like any other enormous spacecraft that we've seen scrolling across the screen, comfortingly familiar against the strangeness of the alien ship.

So there's some plausible evidence behind the concept of the mathematics of desire. There do appear to be regularities, which we can measure objectively, in the appearance of things that people find aesthetically pleasing. One tool that can help us investigate this phenomenon is a mathematical concept called *game theory,* which started out as a way of predicting the likely outcomes of games of chance, such as poker. It has come a long way since then.

Game theory recognizes that you can't predict exactly what will happen in, say, the next hand of a poker game. However, you can predict how *likely* a particular outcome is. For instance, if you hold four queens, there's not much likelihood that the other player's cards will all be stronger than yours, and there's a large likelihood that all of them will be weaker. Over time, the probabilities grind down the occasional flukes, and unsurprisingly, the house always wins in the end.

If you apply game theory to the way humans select their mates, it provides some powerful insights. Let's take height as an example. Tall people tend to be perceived as more attractive and as having higher status than short people. Heterosexual men tend to prefer women who are shorter. So, if you plot out the permutations and simplify the explanation to keep it manageable, a tall woman will have greater perceived status than a shorter woman (useful when competing with other women for job promotion and the like), but a tall woman will be limited in the number of men who find her attractive, compared to a shorter woman. So, in intrasexual (within-gender) competition, tall women are at an advantage relative to shorter women, but in intersexual (between-gender) attractiveness, they're at a disadvantage.

That's obviously a simplification—if you want a more detailed, complex account, the evolutionary ecology literature is crammed with discussion of this topic—but it gets across the underlying point: "attractiveness" isn't a single thing, but rather takes you into the realm of trade-offs and different facets of what is meant by "attractiveness."

Attractiveness, it turns out, is also context-sensitive: for instance, in peaceful times, women tend to prefer less ruggedly masculine faces (e.g., Robert Pattinson), whereas in times of risk and conflict, they tend to prefer more traditionally masculine faces (e.g., Daniel Craig).

Another trade-off in mate selection is between *display* and *cost*. The classic example is the peacock. There have always been obvious questions about how something as outrageously cumbersome as the huge tail of the peacock could help its owner survive. At first glance, it seems a lot more likely that peacocks with large tails would soon be removed from the gene pool by predators, while peacocks with smaller tails and higher maneuverability would survive and spread their less flamboyant genes. Yes, the tail is an encumbrance. But that, Charles Darwin argued, is the whole point. It's saying to the peahens, in effect: "Look at how fit and strong I am—able to thrive despite having a tail as huge as this." In game theory terms, the payoff (attracting more mates and producing more offspring) outweighs the risk of being killed earlier by a predator; over time, the house—in this case, the peacock most attractive to peahens—will win.

There has been a lot of solid research into this phenomenon over the last few decades, and the result that keeps showing up is that the more flamboyant the display, the better the health of the animal doing the display and the better their chances of reproducing and passing on their genes. So there's a strong selective pressure toward more and more flamboyant displays. In the field of evolutionary ecology, this escalation is known as an arms race, which takes us into some of the more uneasy shores of human desire.

Arms Races, Superstimuli, and Spooky Robots

Arms races occur everywhere. The fashion industry is a prime example; if you look through a history of fashions, you'll see arms races where designers compete with each other in the richness of the fabrics, the colors, and the shapes of the fashion models—thinner models in one decade, more voluptuous models in the next.

The same is true of automobiles and mobile phones. You'll also see it in behaviors; when a group feels threatened, it tends to rally round more and more extreme manifestations of its core values—people competing to be the loudest and most extreme demagogues, the most fanatical true believers in religion and politics, the most hard-core of a lifestyle group.

Which raises a lot of obvious questions, such as whether there is a point where arms races hit a ceiling and can't go any further. That question was partially answered in the 1920s by the discovery of the supernormal stimulus, or "superstimulus." Researchers Konrad Lorenz and Niko Tinbergen discovered that the animals they studied would react strongly to an artificial stimulus that exaggerated a feature beyond anything that existed in nature.

For example, a bird would jump at the chance to hatch a fake egg that strongly resembled its own eggs—and was two or three times larger. In fact, animals reacted more strongly to that artificial stimulus—larger eggs, larger tails, more colorful bills, and so on—than to the less extreme real thing. In essence, nature didn't seem to have a built-in "off" switch in most of these cases; you had to push the stimulus far beyond anything natural before it lost its effect.

The fashion industry has been using this principle for centuries, without spelling out or even realizing what it was doing. In terms of superstimuli, cosmetics make the skin smoother than anything found naturally; lipstick does the same for the smoothness, color, and shininess of the lips. Airbrushing and photoshopping remove blemishes from images of models' faces, making them more symmetrical and smoothly textured than any human.

This area has received a lot of attention from researchers. V. S. Ramachandran, an Indian-born neuropsychologist, argued that a lot of art operates by taking a feature and exaggerating it, in a more sophisticated and subtle form of the cartoon. The reaction from the mainstream cultural studies community was predictable. Ramachandran and his colleague William Hirstein weren't actually claiming that the emperor of traditional art criticism was wearing no clothes, but a lot of critics interpreted their work that way. Reaction was hostile, to put it mildly.

That doesn't necessarily mean that the art critics were right in their claims that there was a lot more to art and culture than just a collection of simple perceptual effects. Humans in general tend to perceive many skills, tasks, or products of human ingenuity as more complex than they really are. It's only when you unpack what you're studying and comprehend its deep structure that you realize how elementary it is. You will recall that chess was believed to be far too complex to be tackled by a crude machine, yet actually turned out to be fairly simple to model computationally, at least up to the level of a skilled amateur. The same happened with a lot of other problems, like deciding which cargoes to unload in which sequence at a busy port, or when to fire which engines for how long on a long, complex spaceflight. Humans find such tasks difficult or impossible because they rapidly overload our short-term memory, which can only handle about seven items for a few seconds, and are vulnerable to our bias-prone long-term memory. Computers don't have the same built-in limitations as the human brain. Given the shortcomings of the human cognitive system, maybe human aesthetics really are based on very simple principles.

Ironically, Ramachandran and Hirstein are authorities on another effect, which crops up everywhere and is particularly relevant to the debate about the underlying principles of aesthetics. The effect is called *confabulation,* which involves humans' attempts to rationalize their beliefs—even if what they say in the process sounds completely insane to everyone else. Neurological damage may lead a patient to believe that he has lost, say, his left arm. No matter how hard you try to make the patient see that his arm is still attached to his shoulder, he'll say, no, that isn't my arm; it's someone else's. When you ask the patient how that could possibly happen, the patient responds: "Maybe a medical student left it there as a prank." If you ask why a medical student would do that, the patient swiftly moves onto sensible, solid ground, talking about what medical students are like, and gets safely away from his hard-to-maintain original claim. Confabulated explanations can rapidly become very elaborate, with a high degree of internal consistency, if you don't question their initial (and often totally unreal) initial assumptions.

It's plausible that low-grade confabulation is something we all do every day as a completely normal part of making sense of the world; we don't have all the information we need, so we make reasonable assumptions to fill in the gaps. Why are my car keys on the table instead of in the usual place? I must have left them there when I set down the junk mail that I picked up in the hallway after getting home. That explanation is probably confabulation, but it's not pathological. If that model is true, the alien limb patients are using a perfectly ordinary mental tool, the only difference being that they're using it a lot more vigorously than the rest of us.

This theory has big implications for anyone trying to check the correctness of expert reasoning. It could be very difficult to determine whether an expert's explanation is just a series of confabulations, little different from the troubled patient's belief about his alien limb.

Experts, like everyone else, are wired with aesthetic preferences that almost inevitably produce biases in their thinking—toward symmetry in designs, toward the interesting, and in many other directions, sometimes contradictory, sometimes mutually reinforcing. Architects are predisposed to create beautiful, award-winning designs that will bring them status, much as a peacock's tail does. Aviation engineers think *automatically* in terms of symmetry, even when an asymmetric design might be better. And so on.

Fortunately, there are some clear limits to superstimuli and arms races. One involves physical limitations: automobiles competing in size eventually hit a limiting factor in the form of road width, and fashion models competing in height and small waistlines came up against the limits of human physiology. Another constraining factor emerged, unexpectedly, from Japanese robotics. Japanese researcher Masahiro Mori found that if you made a robot increasingly human-looking, people tended to like it more. Mori also found, however, that there's a point where suddenly the robot starts to produce very uncomfortable responses in people because it has reached a point where it's no longer clearly a robot, nor clearly a human. Mori named this phenomenon "the uncanny valley."

Since then, research has consistently found that there's an awkward gray area in aesthetics that leaves people feeling uncomfortable. Most researchers have focused on the area between humans and nonhumans. It has obvious implications for horror movies, where the monster usually lives very obviously in the uncanny valley and provokes strong feelings—horror, disgust, shock—when it first pops up on the screen.

Some researchers are investigating the impact of the uncanny valley on sociology and anthropology, where human societies and groups define themselves in terms of "us" versus "not us" and are uneasy about any borderline cases. Most societies have traditions about some form of wild man—for instance, a creature that's not really human but also not really animal. Such creatures, which populate folklore, evoke strong reactions of horror, too. One wonders if they are examples of the uncanny valley, used to mark the mental borderland where humans end and animals begin.

One question that doesn't seem to have been picked up by industry is whether this effect is just one example of something much more widespread: whether most people feel uncomfortable with *anything* that doesn't fit neatly into a pigeonhole, regardless of whether that thing is a robot or an automobile or a song. If so, that has big implications for anything being marketed as a "hybrid" or as a "best of both worlds" product. Some of these products, like tablet computers, came out as the best of both worlds. Others didn't.

This will be an issue for next-generation vehicles running on solar power or fuel cells or whatever. Are they falling into an uncanny valley between "automobile" and "bike"? When Segways (two-wheeled personal transporters) were first introduced in 2001, they were mocked by the media as strange creations whose riders were somewhere between pedestrians and cyclists. The device certainly hasn't replaced cars or bikes, as some enthusiastic adopters first predicted, but they have carved out a niche in the tourism industry as a safe way for groups of people to cover ground in areas where tour buses or other means of transportation aren't desirable.

Products like Segways run into completely unintended problems of perception unrelated to actual facts about their performance. The little that is known about the mathematics of desire tells us that we as a species are vulnerable to superstimuli, and that we are put off by the uncanny valley. That could have far-reaching implications for expert reasoning in areas as diverse as medical research and consumer electronics design.

There are some possible examples of arms races leading into uncanny valleys within human sexuality. Some researchers have argued that some paraphilia—fetishes, as most people know them—are actually examples of mainstream sensory biases taken to an extreme. For instance, a lot of paraphilia involve leather and rubber and plastic, all extreme examples of the mainstream human preference for things that are smooth, glossy, and shiny. However, although the "smooth and shiny" explanation is *consistent* with a lot of the evidence, *that doesn't mean it's necessarily true.*

Researchers into evolutionary ecology are well aware of the risks of extrapolating from partial evidence into plausible explanations that bear no relation to reality or, worse, are almost right and therefore mislead researchers for years to come. One good example of why this approach needs to be treated with caution comes from the bowerbird.

Bowerbirds

If you do an online image search for *bowerbird,* you'll find lots of pictures of a bird with glossy black and blue plumage, probably standing on the ground in front of a weird nest-like construction lavishly ornamented with bright blue stones, pieces of shiny blue plastic, and chunks of broken blue glass. This is the bird's bower. Some of these bowers are so big and so complex that when the first explorers found them, deep in the bush of Australia, they didn't believe the bowers could have been made by animals and assumed aboriginal Australians had produced them. But no, they really are made by birds: the male satin bowerbird tempts prospective mates with the size and beauty of the display he has put together.

A blue bird producing a courting area decorated with blue, glossy ornaments looks like a classic case of superstimuli. The fact that that the ornaments are smooth and shiny suggests that, like the preference for symmetry, this is something that runs deep in the animal kingdom.

The idea of innate preferences for smooth, shiny things makes sense. Some features of your appearance can be faked, but others, such as the glossiness of your hair or the smoothness of your skin or the brightness of your eyes, are reasonably honest indicators of how healthy and youthful you are—and both are factors that affect your perceived attractiveness to potential mates. More speculatively, that rule of thumb, that heuristic, can be applied to other survival-related topics. For instance, clear, shiny running water is more likely to be healthy than discolored stagnant water; for species that eat fruit and leaves, a glossy appearance is usually a good indicator that the food is ripe and fresh. It's a plausible story, not least because it fits with the way the natural world tends to reuse tried and tested components for new purposes whenever possible; using the same heuristic for selecting food, water, and mates would make a lot of sense.

Researchers soon became very interested in the satin bowerbird. There were a lot of juicy questions they could try out on it. Researchers wondered if the degree of superstimulation (as indicated by the size and colorfulness of the bower) related to the reproductive success of the bird that made the bower. Did the shininess and smoothness of the ornaments relate to the male bird's reproductive success? If in fact birds with rarer and more exotic ornaments had a better sex life, such a finding might have implications for our understanding of human economics—with expensive sports cars and designer clothing filling in for the bird's ornaments. More tantalizingly for models of economics, did rarer and more exotic ornaments improve reproductive success, acting as the bowerbird equivalent of a shiny expensive sports car and designer clothing? It was a rich area for armchair speculation, and because there are several species of bowerbird, there was scope for some very interesting comparisons across species.

The findings, best expressed by the title of a paper by researchers Joah Madden and Andrew Balmford, couldn't be clearer: "Spotted Bowerbirds *Chlamydera Maculata* Do Not Prefer Rare or Costly Bower Decorations." Other studies told a similar story; whatever was going on among bowerbirds, it wasn't a simple case of conspicuous consumption or of classic supply and demand. Nor was it a simple case of shiny decorations being good; some species don't use shiny decorations. Gerald Borgia found that satin bowerbirds were very particular about choosing ornaments that were blue, but they weren't picky about whether those ornaments reflected ultraviolet light, even though reflected ultraviolet light is an important part of the bird's plumage display. We still don't know what's going on with the bowerbird's choice of home decoration, but the moral of this small story is familiar to researchers: be very wary of big unifying theories that look too glamorously elegant to be true. For every theory that does manage to stand the test of time, there are a lot more that fall by the wayside.

That doesn't mean the core concept behind the mathematics of desire is necessarily wrong. There are a lot of individual elements that still look pretty solid. How to integrate them, though, and what the implications are for human biases and errors, are challenging questions. One relevant issue that hasn't received much attention until recently is what exactly is involved in "hard-wired" preferences, or "instincts." With the growth of computer science, particularly robotics and artificial intelligence, researchers started asking hard questions about just how precisely this mysterious mechanism was supposed to work. Their findings demolish old models of nature versus nurture, reassembling the pieces into something subtly but profoundly different.

An example that has been studied a fair amount is insect locomotion. A lot of robotics researchers have made the reasonable assumption that insects are a good challenge, in terms of simulating the insects' behavior with machines—complex enough to be difficult but simple enough to offer some hope of success. There are quite a few robot insects under development or already in production, though most are a lot bigger than living insects. The researchers building the early robot insects soon learned that it was far from easy to write instructions telling the robots how to move. It was so difficult, in fact, that some researchers simply gave up on that approach and instead equipped their robots with artificial neural networks (ANNs), a technology modeled on the human brain. What they did was, in effect, to tell their robots: "Find out how to move as efficiently as possible. I don't care how you do it; just find the quickest way of getting from A to B."

What happened was interesting. Time after time, with different robots and different researchers and different labs, robots converged on the same method. It's known as the alternating tripod: you keep three feet on the ground while you move the other three feet forward, then you put those feet down and move the three feet that had been stationary. It means that you always have a stable tripod of feet on the ground, so you're unlikely to fall over. It's a widespread pattern among living insects.

If a nineteenth-century zoologist had seen a newly hatched insect going through the same process and ending up with the same result, he would probably have recorded it as a case of "instinct" and practice. But robots don't have instincts; they just have instructions to learn a way of getting around efficiently. They certainly don't have an "instinct" for walking with the alternating tripod gait.

So when you start looking at how the brain actually works, you start having to rethink a lot of old assumptions about culture versus instinct and how we really operate. Research will probably find that a lot of human behavior previously believed to be very complex is actually very simple, once you know the right variables.

Reappraisal

So far, we've just been looking at possible inbuilt aesthetic preferences. There's one big factor in human aesthetics that we haven't yet discussed, which cuts across both inbuilt preferences and human brains looking for easy, predictable solutions. That factor is *novelty*. People get bored easily, so they are attracted to novelty, provided that it's within their preferred comfort zone. Those who unlock the secrets of human novelty could—and have—become very wealthy indeed. Much of the mega-billion-dollar entertainment industry is based on supplying novelty.

One simple way of measuring novelty is a concept called inverse frequency weighting, which measures how information-rich something is. You count how many instances there are of the thing in question—for example, how often a particular word occurs in a particular film script—and then invert it, as 1 over that number. So, say you've found that there are 23 mentions of the word *love* in a particular script, 354 mentions of *gun,* and 3 occurrences of *parenthood.* These figures are then converted into the fractions 1/23, 1/354, and 1/3 respectively.

The inverse frequency value for *parenthood* (1/3) is about seven times as high as the value for *love* (1/23), and about a hundred times higher than the value for *gun* (1/354). From this we can say that the word *parenthood* is the most information-rich of the three within that script, followed moderately closely by *love* and much farther back by *gun*.

And if you're wondering why anyone would care about quantifying novelty, you might want to think about how Google became one of the biggest corporations in history. Every time you do a Google search, the search engine determines the rarity of each keyword. It assumes that the rarer keywords are the more significant ones, with higher information value, and places them on the first page of your results. The search engine thus calculates rarity using a variant of inverse frequency weighting.

There are many other ways of handling information that are equally simple and allow you to measure the novelty of such things as a movie script, a book manuscript, a piece of music, and other works of human ingenuity. One of my students, Nikki Holland, successfully used these techniques to predict likely responses to a webpage.

That has a lot of implications for the entertainment industry, and for such things as the design of product labels and websites. Simply put, if you can measure novelty, complexity, symmetry, and prototypicality, among other things, you can make a pretty good start on measuring what people like and *predicting* what sort of art, books, music, movies, automobiles, or fashion they might like.

If you're a jaded consumer, imagine being able to ask the online music store to find you a piece of music, and being able to tell it whether you want something comfortably familiar or something more challenging, right at the edge of your personal comfort zone. This would be theoretically more powerful than current approaches employed by companies such as Netflix or Amazon, which make recommendations on the basis of "People who bought this also bought that." Such software could revolutionize the way we find items that we didn't know we would love.

Speculations

The mathematics of desire can take us also into uninvestigated areas of human thought and behavior, many of which are heavily politicized. Sociologists and medical researchers often find that a research issue which looks innocuous or highly specialist can turn out to have huge political implications. When sex researcher Alfred Kinsey represented sexual preferences—heterosexual versus homosexual—as a scale, his findings weren't well received in a society that generally viewed sexual preferences as a clean-cut binary distinction between heterosexuality (which was viewed as the norm) and homosexuality (which was generally viewed at the time as a pathological aberration). Think about it: all Kinsey did to arouse such controversy was to visualize the problem differently—with a different form of representation. Treating sexual preferences as a scale rather than a two-way divide was implicitly throwing down the gauntlet to the mainstream values of the time, and it provoked some furious responses.

There are plenty of other examples—in sociology, entertainment, economics, and real estate, for starters—where some apparently obscure representation or research finding is the tip of something much bigger. I'll sketch out a few of these briefly.

Gender politics: Recall the height and status issue we discussed earlier. Researchers who understand the mathematics of desire will want to investigate to what extent women take a double hit as a minority—once for their gender, and again because they tend to be shorter than men. Are they, in other words, suffering from both sexism and heightism? Heightism in general looks to be deeply embedded in human society, but it has not received a lot of attention from researchers. Investigators will no doubt also wonder how many of people's other aesthetic preferences have a dark side, in prejudices so ingrained that we don't even notice them.

Consumer research: Because we know that experts use finer-grained discriminations than novices, on the whole you'd expect novice consumers—children and young adults—to prefer un-subtle products that more experienced consumers find garish and unsubtle. Maybe children's preferences for particular colors and designs are just a by-product of the more general issue of expertise development, rather than involving any hard-wired developmental processes specific to children.

Entertainment: Researchers may well wonder if music and art and drama are not simply ways of regulating our mental arousal levels. What other ways might humans use to fulfill their cravings for novelty? That immediately touches on everything from the early-morning caffeine jolt to the politically charged issue of drug addiction.

How are those issues being studied? Is there some big, unifying way of viewing them that would suddenly make sense of many things that are currently half-understood, the same way the introduction of game theory suddenly made sense of a huge swath of evolutionary ecology? For instance, do people prefer similar uncertainty factors across the outcomes of football games and gambling games and whether the medics in the TV drama manage to find the cure in time in this week's episode of the show? Or do people prefer to have high uncertainty in some areas of their life balanced by low uncertainty in others? Answering questions like that involves bringing together topics that are usually studied separately, usually using different methods in each field; it's a very different approach from how things are done at present.

Another question: how would these effects interact with differences in sensory issues within the population? Do differences in the low-level processing mechanisms in people's brains—things like information-processing rates and short-term memory capacity and vestibular feedback, things that few people except expert neurologists have ever heard of—influence significant differences in the individual's preferences for arts and music and sports and entertainment and lifestyle?

Politics: With some techniques borrowed from the field of software development, you have a shot at modeling cause and effect. If you were to apply that to areas such as politics, you could test the accuracy of expert reasoning in policy decisions. When you also bring in concepts like pattern matching versus sequential processing to tackle that question, it raises some fascinating speculations about political rhetoric. Suddenly, you have the tools to wonder if vote-seeking politicians are simply churning out a word salad that looks good to our pattern-matching circuitry because it has the right words in it. You can also theoretically measure if politicians are getting away with using incoherent and self-contradictory statements. If our brains wrongly decide via pattern matching that everything's okay, they won't bother to use slower, harder sequential reasoning to assess whether the politician's statement actually makes any sense. Would getting policy makers and stakeholders to map out their ideas as a cause-and-effect diagram help make politics and decision making less divisive and more effective?

Sales: If we could create an integrated research approach to human desire, we'd be able to see patterns and deep structures that the experts hadn't thought to look for before. There's an effect that salespeople know and loathe in a wide range of fields. For real estate agents, it takes the form of the client who spends months looking at nothing but lovingly maintained old houses, and then suddenly decides to buy an ultramodern apartment. For music, it's the aficionado whose collection has severely minimalist Philip Glass albums next to punk. How can we possibly hope to understand such a buyer?

One possible explanation is that there isn't actually a contradiction, if you visualize things differently. Suppose you represent music as a circle, with mainstream pop at its center. Now put other types of music on the circle, with their positions showing how far from pop they are. Rock music would be a moderate distance away in one direction, punk further away in roughly the same direction, and Philip Glass way off in the other direction. You get a diagram something like figure 30.

There's a new regularity that shows up in this representation. The Philip Glass music is about the same distance from pop as punk music is. You could do the same with real estate, and you'd probably put the authentic old house about the same distance from the average house as the ultramodern apartment is. Maybe what's going on is that it's the distance—the rarity factor and the novelty factor—that's important to these people. Maybe the distance is more important than the direction. The tools to measure those distances already exist—another of my students, Giselle Martine, used card sorts to do just this type of analysis.

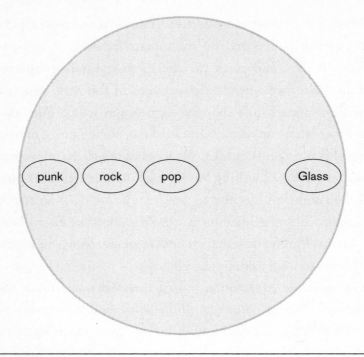

punk rock pop Glass

FIGURE 30

Warfare: One could analyze Cannae and other battles in light of what we suspect about the mathematics of desire. Military historians have been fascinated by Cannae down through the millennia. Other battles have arguably had a much bigger effect on history or were much more dramatic in their outcome, but those don't usually get anywhere near as much attention. The battle of Gaugamela, for instance, saw Alexander the Great take his cavalry in a unique move across the front of the Persian army of the emperor Darius, then strike to the heart of that army in a devastating blow that won the battle and made Alexander ruler of Persia. Cannae, in contrast, was just another battle that Hannibal won during a war he eventually lost. Yet Cannae is treated with far more reverence.

When you describe the two battles in terms of the mathematics of desire, the differences become apparent, and you start to see why one might attract what is almost a cult following (figure 31). Cannae ticks all the boxes in terms of the mathematics of desire. It has symmetry, familiarity, one moment of novelty when the Romans realized that they were seeing the wrong familiar pattern, and at the end complete closure in the killing of the Roman army. Gaugamela, in contrast, ticks none of those boxes.

That same set of contrasts was to emerge over two millennia later, in a series of decisions that changed the course of modern history. When World War I began, the Germans had a massively detailed plan for their attack, named the Schlieffen Plan, after its creator. It involved a deliberately asymmetric attack. The left wing of the German forces would fight a holding action, keeping the French from counterinvading Germany. The much stronger right wing of the German forces would meanwhile hook around through Belgium, crushing the French forces from that flank in a swift victory. What happened next has always intrigued historians.

The Germans were well aware of the reasons for using an asymmetric plan; they had worked out in minute detail how

Cannae	Gaugamela
Symmetrical layout	Asymmetric layout
Initially followed a familiar pattern	Initially followed an unfamiliar pattern
Switched to a different familiar pattern	Switched to a different unfamiliar pattern
Closure: enemy annihilated	No closure: most of enemy survived

FIGURE 31

many troops would be needed on the right wing, and they knew that they didn't have enough resources to try for another Cannae with a double envelopment. They knew it, but somehow they weren't able to resist the temptation to move troops from the right wing to the left wing, trying for that symmetrical double envelopment anyway. It failed, and the war turned into the long, bloody ordeal that the combatants had wanted to avoid at all costs.

Ascribing the change of plan to a simple desire like symmetry may sound far-fetched, but plans are made by human experts, who need to have an overview in their minds. Those plans will be affected by the limitations and biases of the human brain. A bias from a desire for symmetry is no more unlikely than a bias from "strong but wrong" errors, and those errors are completely accepted by safety-critical systems research as potential root causes of disasters. Perhaps, ironically, the mathematics of desire itself falls into an uncanny valley as a discipline—neither aesthetics nor psychology nor mathematics, but a hybrid of all three.

Because the mathematics of desire is so deeply woven throughout our world, it's difficult to step back and see just how pervasive it is. The commonalities are usually obscured by the way our society compartmentalizes reality. We think of sports, music, and action movies as three very different categories, and that makes it harder for us to spot the underlying commonality: the realization that sports, music, and action movies are all about a trade-off between predictability (rules, conventions) and novelty. We think of *the plot in fiction* as a separate category from *crime,* and lose sight of how important the desire for closure is in both arenas. A lot of scientific research is also about that same quest for closure.

That's a feeling I know all too well. With the Voynich case study, the rational part of my brain knows that the hoax explanation is the best fit for the evidence, but the part of my brain driven by the mathematics of desire keeps niggling away, wanting a clean closure with definitive proof of who carried out the hoax, and when, why, and how. In the real world, though, you don't always get closure. That's one reason fiction is so popular; it's tidier and more satisfying than reality. That's fine when you know you're dealing with fiction, but it gets scary when the line between fiction and reality gets blurred, somewhere on the spectrum between conspiracy theory and whacky ideologies and legitimate personal opinion.

The desire for order in the world is a theme that features prominently in our second case study, on autism.

Fuzzing
into Mush

THE VOYNICH CASE STUDY HAD BEEN TANTALIZING— it suggested that our Verifier method could be used fast and light and cheap, completely the opposite of what we had expected. With those experiences fresh in our minds, we decided to apply Verifier to a major unsolved problem in medicine and see what new insights we could uncover. Sue Gerrard took up the next case study. The subject was autism.

From the beginning, autism seemed like a perfect unsolved case for Verifier. The disorder has proved a tough nut to crack, despite a great deal of research over a long period of time. Also, it was still an unsolved problem even though there appeared to be plenty of data and generally accepted definitions of the key terms. It didn't look like one of those problems where researchers lacked the information they needed or hadn't yet agreed on the key methods and concepts needed. To us, that suggested there might be some deep, subtly flawed assumption at the heart of the standard model in autism. Though Sue followed essentially the same Verifier steps that I had followed with the Voynich Manuscript, autism was a much bigger problem. She found that the surface appearance of consensus about key concepts was misleading. Once she started to investigate the deep structure, a very different picture emerged, one that was very different from the picture you find in most articles and documentaries about autism.

We'll start with the definitions. The most widely accepted definitions of psychiatric disorders are listed in two regularly updated publications—the *Diagnostic and Statistical Manual of Mental Disorders (DSM)* and the World Health Organization's *International Classification of Diseases (ICD)*. These define autistic disorder as *impairments in social interaction and communication* together with *restricted repetitive and stereotyped behaviors*. That second phrase refers to behaviors that are within a narrow range—there's little flexibility—and that tend to follow repetitive patterns. A child might spend long periods every day lining up toys, or might repeatedly flap her hands.

The landmark paper first proposing that autism was a medical condition in its own right was published in 1943 by Leo Kanner, an Austrian psychiatrist who had emigrated to the United States in the 1920s. During his work at Johns Hopkins University Hospital, he came across children whose unusual behavior didn't seem to match any developmental disorder recognized at the time. Kanner's paper included case studies for eleven children— eight boys and three girls—that he felt showed symptoms of a syndrome that had never been described before. The following year, another Austrian, a pediatrician named Hans Asperger, published a paper describing four boys he had seen at the University Paediatric Clinic in Vienna. Asperger described a syndrome—now named for him—whose symptoms were similar to those described by Kanner. Kanner and Asperger's use of the term *autistic* wasn't a coincidence. *Autism* was a term that had been coined decades earlier by the Swiss psychiatrist Paul Eugen Bleuler to describe the withdrawn state seen in schizophrenic patients.

Several of the children described by Kanner had been diagnosed with schizophrenia or hearing impairment. Kanner felt these diagnoses did not adequately explain the symptoms and that there was a distinct and previously unreported syndrome involved. Asperger reached the same conclusion about the group he observed. Children in the two groups were similar in a number of ways, most obviously because they showed little interest in other people and had problems with language and nonverbal communication, had unusual responses to sensory stimuli, displayed remarkable abilities such as facility with numbers and excellent rote memory, and shared what Kanner called "an insistence on sameness"—they were highly distressed by broken objects, disrupted routines, or changes in diet. However, there were also many differences among the children. Some were mute; others could speak well but used pronouns oddly or spoke in an unusual tone of voice. Many were clumsy, others very dextrous in specific areas. Some had good cognitive skills; others didn't.

For some years afterward, autism was seen as a relatively rare, complex, and enigmatic developmental disorder. Asperger's work was largely unknown until his paper was translated into English in 1991. Autism came to public attention largely through the publication, in 1967, of child psychologist Bruno Bettelheim's book, *The Empty Fortress: Infantile Autism and the Birth of Self.* Bettelheim supported the suggestion, made by Kanner at one point, that autism was caused by what came to be known as "refrigerator mothers"—mothers who were emotionally cold toward their children. After this assessment, autism became a controversial diagnosis and issue. It implied that people were "making" their kids autistic.

An important step forward came in 1979, when British researchers Lorna Wing and Judith Gould put Kanner's criteria to the test and found that children who met his criteria showed three co-occurring impairments: in social interaction, communication, and imagination. Lorna Wing developed the concept of autism as a "spectrum"—a range of conditions linked either by symptoms or by cause. Wing and Gould described their model as a "triad of impairments." Both concepts have been influential. The concept of a triad of impairments shaped the definitions of autism in the DSM and the ICD. Since then, there has been a huge amount of research into possible causes of autism, but when you ask the simple question "What *is* autism?" the answers in the scientific literature all take you back to a syndrome, a triad of impairments, and a spectrum.

It was clear to us that if a field has trouble even defining what it is about, then it is probably a good case for Verifier.

Medical Models

At this point it might be helpful to take a closer look at some of the models used in medical diagnosis. Kanner and Asperger each suggested that the common characteristics of the children they described formed a *syndrome*. A syndrome is a group of features, signs, or symptoms that often occur together.

In Lorna Wing's *spectrum* model, the symptoms vary in degree and in kind. Although children diagnosed with autism have in common the three core features that define autism, different children often have different impairments, which also vary in severity.

All three autism models—the syndrome model, the triad of impairments model, and the spectrum model—deal with signs and symptoms; signs and symptoms are all they can deal with, because we have little information about the causes of autistic characteristics. So a diagnosis of autism isn't a diagnosis in the same sense that a diagnosis of measles is a diagnosis—where, if necessary, you could haul out a microscope and point to the precise strain of *Morbillivirus* causing the illness.

All three models look perfectly reasonable at first glance. There is no doubt that some children do show the features that currently define autism—impairments in social interaction and communication in addition to restricted, repetitive, and stereotyped behaviors. Those children would qualify for a diagnosis of autism, but this doesn't mean that autistic characteristics must be caused by a distinct, clearly defined disorder or a group of related disorders.

Look at it this way. Children with a sore throat, fever, and skin rash were never diagnosed with "sore throat fever skin rash syndrome." This was because this collection of symptoms was so common that parents and doctors quickly noticed subtle similarities and differences among groups of patients and were able to reliably identify the symptoms of different diseases, such as measles, chicken pox, or scarlet fever, many years before the causes of the diseases were known. In other words, not everybody with a sore throat, fever, and skin rash has the same disease. Nor are a sore throat, fever, and skin rash necessarily all caused by the same disease in an individual patient. Each symptom might have a separate cause. Which is why a good doctor will look very carefully at the individual patient's symptoms.

Doctors were able to distinguish fairly accurately between common childhood illnesses using a precise definition of the symptoms. Did the skin rash consist of reddening of the skin, a mass of small red spots, or blisters? What did the tongue look like? The throat? Was there a cough, or conjunctivitis? The doctors' focus was not on the sore throat, fever, and skin rash per se—although symptoms described at that level give a pretty good idea of what illnesses could be involved—but on the *type* of sore throat, fever, and skin rash, and on the other symptoms associated with them.

A clue as to why the causes of autism might be proving so elusive may be gleaned from descriptions of the types of behavior that are impaired in autism. The names of the behavior—social interaction, communication, repetitive behavior—all looked easy to understand, even for a nonspecialist, a triumph of clear, plain English over jargon. Unfortunately, clear, plain English can often be a recipe for trouble.

The Attractions of Being Boring

Once I was teaching a strong batch of students who were doing significantly worse than expected on their assessed written coursework. Although they were clearly familiar with the technical concepts they had been taught, this wasn't coming across in their answers, and that dragged down their marks on assessed work. To find out why, we used our old friend, laddering, the technique designed to unpack concepts systematically until you reach the equivalent of bedrock.

We started by asking the students, "How can you tell that something's well written?"

Each student came back with a list of features of good writing, including something along the lines of "well presented."

Next we asked, "How can you tell that something's well presented?"

Again, the replies came back fairly consistently: features like the text being spell-checked, neatly formatted, and interesting to read.

Drilling down to the third level, we asked, "How can you tell that something is interesting to read?"

A recurrent theme in the answers: it flows easily.

And how can you tell that it flows easily? Answer: *because it isn't cluttered with lots of references and technical terms.*

That made a horrible kind of sense. We now knew why the students were throwing points away every time they were assessed. When you ask experienced academics the same questions, they start off with quite similar answers to the students, as far as "well presented." Then everything starts to head off in a very different direction. The academics are very, very concerned about whether the text demonstrates that the writer has a thorough, masterly grasp of the subject . . . by using lots of references and technical terms.

Exactly the opposite of what the students were doing.

So the students weren't forgetting what they had learned in the course when they did assessed work; instead, they were *deliberately* filtering it out, in the mistaken belief that this would get them higher marks because their writing would flow more easily. Since they were a particularly bright group, they were doing a very good job of throwing away the very things that would have brought them good marks.

We addressed that immediately with lectures and workshops where we explained just why we viewed the technical terms and references as being so important, and why those took priority over flow and interestingness in assessed work. For other purposes, like writing broad-readership magazine articles or a book like this one, the priorities would be very different, for very sensible reasons, but in assessed work, the priority is to show you've got the knowledge that is being assessed.

You don't always get the same patterns when you drill down. In some disciplines, you get tight, precisely defined terms at each new level. You get the same terms from all experts at each new level, and it all bottoms out in clearly defined terms that you can measure or observe in precise detail.

In other disciplines, though, you find that the answers soon start "fuzzing out into mush." You don't need to do laddering on a human expert to find out whether you're dealing with a tight discipline or a fuzzy one; you can just follow the strands in the literature, looking up the technical terms in specialist dictionaries or textbooks, then looking up the explanations of those terms, and the explanations of those explanations, and so on until you reach bedrock. In tight disciplines like chemistry or geology or neurophysiology, everything from journal articles to textbooks will use terms that bottom out unambiguously into definitions you can quantify or could nail down in a lab experiment if you had to. If the terms don't do this, then you're dealing with one of those disciplines that fuzzes out into mush.

As you might suspect, drilling down into concepts like "social interaction" and "communication" and "restricted repetitive and stereotyped behaviors"—the very concepts that supposedly defined autism—swiftly fuzzed out into mush.

Groups Versus Individuals

When you start looking at how the experts define concepts such as "social interaction," "communication," and "restricted repetitive and stereotyped behaviors," you find that different authors propose different interpretations. When you chase down the definitions of *those* terms, they in turn fuzz out, usually taking it for granted that all readers would interpret the lowest-level terms in the same way, just as we had taken it for granted that the students would interpret "good presentation" the same way we did.

Having different interpretations in a discipline isn't necessarily a problem; it can actually be an advantage in some situations. But that wasn't what Sue was seeing as she applied Verifier to the autism literature. When an autism researcher used the terms "social interaction," "communication" and "restricted repetitive and stereotyped behavior," it was not clear in the literature that everyone had exactly the same interpretation of how those terms were being used. The children and adults being described clearly did have problems with something that you or I would probably call "social interaction." But that phrasing is not nearly as precise as the way doctors talk about, say, the medical textbook definition of a popliteal aneurysm or Dupuytren's Contracture. It's like the difference in precision between "broken leg" and "greenstick fracture of the tibia."

The reason "social interaction," "communication," and "restricted repetitive and stereotyped behavior" fuzzed out into mush is not that different experts have subtly different perceptions of what is meant by those terms—although it's likely they do—but because each term encompasses a huge range of widely differing behaviors. There's far more variation in each of these categories than there is in categories such as "sore throat," "fever," and "skin rash."

In autism, the definitions are so broad that a child who appears to be completely unaware of other people, cannot speak, and spends all day rocking back and forth could qualify for the same diagnosis as a child who avoids eye contact, talks endlessly about his own interests, and spends all day lining up toy cars. These children certainly have symptoms in common at one level, but at another level their symptoms are totally different. And no one knows if they have the same underlying disorder, or if they have related but different disorders, or unrelated and different disorders.

All this might seem like pedantic hair-splitting. But the autism models in use have serious implications for three very significant issues: how children are diagnosed, how researchers design their studies, and in most Western nations, at least, what type of funding is allocated to the children's families or to investigation of autism by funding agencies.

Autism researchers are generally well aware of the differences among individuals diagnosed with autism and try to take account of this in their research designs. Research participants are carefully matched on factors such as IQ or developmental age or language ability. Researchers make sure that autistic participants all have a formal diagnosis of autism using the same diagnostic criteria—and that's where the problem lies. Because the criteria for diagnosis are so fuzzy, with each category encompassing a huge range of specific behaviors, it's possible that participants in an experiment could qualify for a diagnosis of autism on the basis of completely different behavioral impairments and completely different causes for those impairments. So, although those participants would all have a diagnosis of autism, that wouldn't mean that they were similar when you got down to the specific impairments that were affecting them. That has far-reaching implications for the ensuing research.

It's as if we have two standards in autism research—one that applies to the group and another that applies to the individual. Many autism researchers start from the group features, breaking them down to see if specific aspects, such as impairments in cognitive executive function or theory of mind, might be involved. But since these concepts also refer to a large number of behaviors, and different children show different impairments in executive function or theory of mind, research into these hypotheses has been sidelined. The research findings tend to come up with inconclusive or contradictory findings because they don't always explain the autistic features of the whole group.

Sue wondered what would happen if you ignored the group characteristics and focused instead on the characteristics of individuals. So she tabulated Kanner's observations of the eleven children in his case studies. Kanner collected quite a lot of information about those children: their family history, birth, development, cognitive and motor skills, as well as their social interaction and communication. Putting Kanner's data into a table meant that similarities and differences could be seen at a glance. Kanner's records weren't completely systematic, and Sue spotted substantial gaps in his data. This was not necessarily an oversight on his part—sometimes information about patients just isn't available. Even so, it was clear that as well as disruptions in affect, the children described in his study had a number of other similarities that had attracted less attention from researchers.

Sue wondered how easy it would be to track down the possible causes of autism by investigating those other similarities, which are less complex than social interaction or communication. When you do that, you find a range of possible physical explanations relating to well-established medical conditions. For example, a commonly reported and relatively simple feature of autism is impairment of "eye-to-eye gaze." This is supposed to mean that an autistic child has trouble making eye contact. However, even "eye-to-eye gaze" turns out to be complex when you start to unpack it. There are numerous possible causes for a child avoiding eye-to-eye gaze, each with different implications. A child may simply not have learned social conventions about when eye contact is appropriate. Or he may be physically uncomfortable making eye contact. Some autistic people experience "visual hyperacuity," which means their eyes are so sensitive that they find it uncomfortable to look directly at anything shiny—like other eyes.

Another possible reason for an impairment in "eye-to-eye gaze" is not visual at all, but, strangely, *auditory*. Most people watch the other person's eyes very attentively when they are having a conversation. Some autistic people have been found to pay more attention to the mouths of people who are speaking. Since speech and language problems are commonplace in autism, a likely explanation for watching someone's mouth is that it provides visual information in addition to the auditory information gleaned from speech. This is a bit like people who are hard of hearing using lipreading, except that researchers have found that most people with normal hearing also use *visual* cues to help them discriminate between speech sounds that sound similar.

So not only do apparently self-explanatory ideas such as "social interaction" turn out to be extremely complex, but what looks at first glance like a simple low-level feature of autism—

impairments in "eye-to-eye gaze"—turns out to be pretty complex as well. But with one major difference: the science there, grounded as it is in human physiology, is not fuzzy.

If you really want to get serious about diagnosing an autistic person's vision issues, you can measure how far offset from center the person's gaze is. You can measure the density of their cone cells—the cells in human retinas that help us see fine detail and changes in images—and the distribution of cones responding to different wavelengths of light. You can measure how much time they spend looking at the mouths of speakers and whether they have problems discriminating between speech sounds if they don't have visual information from lip movements. You could check whether *all* autistic people have impairments in "eye-to-eye gaze" for the same reason. You could find out how many people who *aren't* autistic have impairments in "eye-to-eye gaze," and if so, what other features they might have in common with autistic people. In short, looking at the low-level features of *individuals* instead of the high-level features of the *group* has the potential to yield significant amounts of information about what might be causing their autistic characteristics.

There is no question that children with autistic characteristics do have behavioral impairments, but a lot of other associated symptoms have been reported. These include difficulties with eating; digestive problems, including recurring abdominal pain, bloating, and constipation; difficulties with toilet training; bedwetting; poor motor control; impaired mobility; hypermobile joints; allergies; and abnormal reactions to a variety of sensory stimuli. A psychiatrist of Kanner's generation would have been quite likely to relate difficulties with breastfeeding or constipation to disruptions in the oral and anal stages of psychological development, rather than to problems with muscle control or a dietary intolerance. Seventy years on, it's more difficult to justify viewing the physical symptoms associated with autistic behaviors as coincidental individual differences, or as inconclusive or contradictory research findings and therefore not helpful in locating the cause of autism.

The well-known autism researchers Christopher Gillberg and Mary Coleman once listed all the medical conditions associated in various cases with a clinical diagnosis of autism. The conditions include abnormalities in all but three chromosomes, plus six metabolic disorders, one bacterial and four viral infections, and twenty other medical syndromes. Now, there are at least three ways that the medical conditions could be related to the autism. It's possible that the children involved could have had medical conditions that *just happened to occur together with autism*. Or that whatever caused their autism *could have also caused or made them susceptible to the other medical condition*. Or that the medical condition *caused the autistic behaviors*.

Given the wide range of behavioral impairments that could lead to a diagnosis of autism, the third option is a strong candidate. And yet, despite clear evidence that autistic characteristics are strongly associated with physical symptoms, autism is still classified as a *mental* disorder. You have to ask whether this is a helpful classification, or whether it might actually be counterproductive. And classifications can be a political minefield, in areas as apparently different as soccer studies and battlefield medicine.

Looking at the Big Picture

Back in the 1970s, politicians and the media were seriously concerned about violence among European soccer fans in general, and British fans in particular. Sociologists were encouraged via research grants and the prospect of fame to find out why there had been a sudden, unprecedented upsurge in soccer hooliganism. The sociologists took the money, reported back, and ducked.

They almost certainly knew what they were going to find even before they started, but being able to hire staff to do a big, in-depth study would be too tempting to resist. Their conclusion was that compared to what had gone on in the past, 1970s soccer violence was actually pretty tame. Politically, that finding went down like a supercavitating lead balloon. When there's a media panic going on about some popular demon figure, the last thing the media and the politicians want to be told is that they're making a big fuss about something small.

In science, as in anything else, it always helps to pull back and get the big picture, and part of Verifier involves looking systematically to see whether there are any big-picture issues that you need to factor in. One big-picture issue in autism research is a practical point. It wasn't being mentioned much in the popular press, but it was getting a lot of attention in the grassroots support groups for parents of children with autism. The way the medical system is structured has serious consequences in all sorts of ways, as those parents were all too well aware. The types of treatment and support they could access for their children were highly dependent on the formal medical diagnosis. If your child has diagnosis X, a particular treatment is indicated; if the child has diagnosis Y, the recommended treatment is different. A similar process happens with support. Diagnosis X might give you access to payments for supporting infrastructure or access to an educational program; diagnosis Y might not.

So the diagnostic system is a big issue. In the United States, the standard classification of a whole host of psychiatric, developmental, and behavioral disorders is set out in the American Psychiatric Association's manual, the DSM-IV. If a doctor can match a patient's symptoms to a condition listed in the DSM-IV, the patient gets a *formal* diagnosis. That opens doors to support and treatment of various kinds—medical, welfare, education, and so on. If the doctor can't match the symptoms to something in the DSM-IV, there's a risk that the patient will be left out in the cold, ineligible for those forms of aid.

For some conditions, the decision is straightforward. Popular books on science describe cases like the man who mistook his wife for a hat (in the 1985 book of the same name by Oliver Sacks) or people who insisted that their own arm actually belonged to someone else and had been left in the hospital bed next to them (as in V. S. Ramachandran's *Phantoms of the Mind*). Those are clearly cases where there's something wrong with the unfortunate patient. But in other cases, deciding whether a patient's problem should be categorized as a medical condition at all becomes highly controversial.

In World War I, a lot of soldiers were cracking up mentally from the horrors of frontline warfare in the trenches. In the past, they would have simply been shot as malingerers or cowards. Before long, though, it became clear that something else was going on. Some unsung genius invented the term *shell shock*. This medical pigeonhole saved a lot of men from the firing squad and an even larger number of men from the social stigma associated with mental conditions—*shell shock* sounded like an obviously physical condition caused by the human body being shaken by explosions, not a mental shortcoming on the patient's part.

So, if a condition is defined in the DSM, there can be far-reaching implications for everyone affected (not just the patients, but also families and friends and caregivers—there's a big ripple effect). But deciding whether or not a condition actually exists, and what its symptoms are, is harder than you might think.

The Classification of Mental Disorders

What starts off as a working model for a possible disorder can all too easily take on a life of its own. This is known as *reification,* whereby an abstract or hypothetical concept is increasingly assumed to have a concrete existence, often without sufficient justification.

In the case of autism, Kanner was pretty sure he had identified a previously unreported syndrome, but he was still tentative—he made it clear that further work needed to be done. Further work has been done, of course, but to a large extent it has assumed that the *form* of Kanner's explanation for autistic characteristics—identifying common features in a group of children with very different symptoms—and the *content* of his explanation (disruptions in affect) were correct.

We still hadn't found an answer to our simple question, "What *is* autism?" But in her Verifier analysis Sue found some reasons *why* that question doesn't yet have an answer. We've seen that autism is classified as a mental disorder, a distinctly unhelpful designation; it sidelines physical symptoms that could explain many behaviors in autism as consequences of issues such as sensory problems. We've seen that autism is currently *defined* solely in terms of behavior (although proposed changes in DSM-V—due to be published in 2013—recognize that sensory abnormalities might be involved). And above all, we've also seen that the defining characteristics of autism are couched in very *general* terms. Indeed, the defining characteristics of autism have to be couched in general terms because individual characteristics vary widely. But the general symptoms of a group are often not specific enough to track down what's causing the symptoms in an individual patient.

Our Verifier analysis of autism showed that the fundamental problem for autism research is that although the diagnostic criteria for autism do not assume *explicitly* that all autistic characteristics have the same cause, by giving people with very different symptoms the same diagnosis, that's what they assumed *implicitly.*

It's quite possible that the biggest obstacle to finding the causes of autism is not that it's a highly complex and difficult disorder to untangle, but that it might not actually be *a single* disorder or even a related group of disorders. Autistic characteristics could instead be the outcomes of disruptions to physiology and development resulting from a range of factors that are neither mental nor behavioral but very physical indeed, and whose physical symptoms have been in plain sight for the past seventy years.

To sum it up in one simplified analogy, just because a lot of people exhibit shy behavior, that doesn't mean those people all have a mysterious medical condition causing shy behavior. If you restrict the word *autistic* to its original meaning of "self-focused," then saying that a group of children all show "self-focused" behaviors doesn't immediately make you think that they must all have Self-Focused Behavior Syndrome, especially if those children don't appear to have much in common beyond various forms of self-focused behaviors. Instead, if you notice that one child has hearing problems that make conversation difficult, you might start to wonder whether the hearing problems might be causing the self-focus; similarly, if a child has visual problems that make interacting with other people difficult, those problems might be the cause of that child's self-focused behaviors.

At one level, this is saying nothing new. Debates about the classification of mental disorders and the pitfalls of applying the medical model to human behavior have raged for decades. At another level, though, the Verifier analysis of the literature showed clearly that not everyone engaged in autism research— or in writing about autism or working with autistic children— is aware of the debates about classification or theoretical models. For many people, the fact that a syndrome has been formally defined and there is widespread agreement that those symptoms do indeed occur together in some children means that it's safe to assume that some children have a distinct disorder that we call autism, even if it might involve individual differences.

It looked like we had a lot of really interesting results and im- plications for researchers as a result of our Verifier study. Before we get to those, we'll briefly look at two other cases, both of which have a long and controversial history involving disputes over classifications and definitions. Sue's work on autism forced her to make some unexpected detours into these classic medi- cal mysteries. They are dyslexia and chronic fatigue syndrome.

Dyslexia

Dyslexia, like *autism,* started as a descriptive term. It means "dif- ficulty with reading" and was originally used to describe read- ing difficulties in brain-damaged patients who, prior to their stroke or accident, had been able to read without a problem. As state education systems were rolled out during the nineteenth century, it became clear that some children, for no obvious reason, had similar problems learning to read, so a distinction was made between *acquired* dyslexia and *developmental* dyslexia.

Because damage to specific areas of the brain was known to be a cause of acquired dyslexia, it seemed reasonable to assume that children with developmental dyslexia might have damage to the same brain areas. Although that's possible, children's brains are still developing, and there can be lots of different reasons why their brains don't develop normally or why they don't learn a skill normally. In other words, although the DSM definition might result in lots of children being diagnosed with dyslexia, it doesn't mean that all children diagnosed with dyslexia have the same symptoms. Nor does it mean that similar symptoms must have the same cause. It also doesn't mean that symptoms seen in individuals but not in the group as a whole can be overlooked because those symptoms aren't features of dyslexia.

Again, this might seem like hair-splitting, until you look at the implications. If all children diagnosed with dyslexia are assumed to have the same disorder, there's a risk that reading difficulties needing different types of support might be lumped together and treated the same. If dyslexia is assumed to be a problem with the "wiring" of the brain, as is often assumed, problems with eyes and ears or basic information processing might be ignored. And if a child has symptoms not shared by the whole group of children diagnosed with dyslexia, these symptoms might be overlooked as unrelated to the dyslexia, even though they could be important clues for what is causing the problems in that specific child.

Research into the causes of developmental dyslexia has been

going on for the best part of a century, but it remains a subject of intense debate and controversy. To make matters worse, research findings are often inconclusive or contradictory. Early research pointed to visual abnormalities as a possible cause of dyslexia. Recent research has indicated that many children diagnosed with dyslexia have auditory processing problems. This doesn't mean we once thought that "dyslexia" was caused by visual problems but now we know it is caused by auditory problems. Nor does it mean that all children diagnosed with dyslexia have auditory processing problems, nor that all those children have the same auditory processing problems. Ears, like eyes, are complex organs, and there's a lot that can go wrong with them and with the information they process.

Based on our Verifier analysis of some key papers in dyslexia, it looks suspiciously as if dyslexia is treading the same path as autism: that referring to dyslexia and autism as discrete disorders and providing criteria by which each can be diagnosed *reifies* them. They stop being a group of symptoms that often co-occur, and instead are perceived by researchers as having a concrete existence as distinct disorders—*even though there might not be a distinct disorder there.* Again, an analogy might help. Suppose that instead of using the term *dyslexia* you use the term *reading disorders.* If someone told you they were looking for the cause of reading disorders, you'd probably be surprised and wonder why anyone would expect there to be a single underlying cause for all reading disorders. However, if someone tells you that they're looking for the cause of *dyslexia,* that doesn't sound immediately odd. Reification means that any individual characteristics not applying to the whole group of people diagnosed with the disorder can all too easily be marginalized—in research, in individual diagnosis, and in treatment.

Our next example of problems in medical definition is a condition known by various names—the names in themselves are hot areas of controversy. The name we'll use here is *chronic fatigue syndrome.*

Chronic Fatigue Syndrome

During the summer and autumn of 1955, a mystery illness in West London forced the closure of the Royal Free Hospital because so many members of the staff were affected. There were several other outbreaks of an illness with similar symptoms reported in the same year in other places. A distinctive feature of the illness was that it was often followed by persistent muscle fatigue, lasting for many years in some cases.

The illness was eventually classified as a disease of the nervous system and named benign myalgic encephalomyelitis (ME). As time went by, there was increasing debate about the diagnosis of patients with similar but not identical symptoms. So in 1988, the experts decided to go with the "working case definition" we all now know: chronic fatigue syndrome (CFS). Since the syndrome of symptoms often occurs following a viral infection, it's also sometimes called postviral fatigue syndrome (PVFS).

The controversy over ME/CFS/PVFS centers around the fact that the symptoms are often purely subjective. The patient reports them, but they can't be verified by clinical tests. Often nothing shows up in clinical tests. Patients feel tired and weak, have sore throats and painful muscles, can't think clearly, and can't sleep properly, but there are no lumps, bumps, raised temperatures, or abnormal blood tests to give a clue as to what might be going on. So although it's not classified as a mental disorder, you can understand why chronic fatigue syndrome has often been suspected of being one. The outbreak at the Royal Free Hospital has been attributed to mass hysteria, patients are still pejoratively described as suffering from "yuppie flu," and the current recommended treatment is merely management of the symptoms. That's about all the medical profession can offer those affected. This disorder, like autism and dyslexia, is a bona fide mystery.

CFS is also, like autism, an "exclusion disorder." When doctors have eliminated all other possible diagnoses matching the patient's symptoms, they conclude that the patient has CFS. The Centers for Disease Control and Prevention describe the diagnostic criteria for CFS in terms of unexplained, persistent fatigue that results in a significant reduction in activity, together with a group of symptoms, including sore throat, enlarged lymph nodes, and disrupted sleep, that could have a wide range of causes.

The two interventions regarded as having the clearest evidence of benefit are cognitive behavioral therapy and graded exercise therapy. This is hardly surprising, since variations of both have been used for many years to help patients manage the symptoms of chronic illnesses. But not all patients improve with these forms of therapy. Some do, some don't. Some actually get worse.

The science on this syndrome is similarly variable. The experts cannot agree on whether the condition is primarily psychological or physiological.

The three syndromes we've looked at—autism, dyslexia, and chronic fatigue syndrome—have features in common. Each syndrome has a broadly phrased definition, to take into account considerable variation among individuals. Each has proved resistant to decades of attempts to find a cause. Each has a controversial history; many research findings have been contradictory or inconclusive, and the very existence of the syndrome has been disputed. The question hanging over autism, dyslexia, and CFS is whether their apparent intractability is due not to the complexity of the condition but to the way symptoms have been categorized.

Giving a definition and a name to a group of symptoms that often co-occur can have positive outcomes, as in the example of shell shock. Patients and their families are relieved when it's agreed that they are not imagining their symptoms. Having a syndrome with a name can put you in touch with other people with the same syndrome. Sometimes having a diagnosis can get you financial, medical, or educational support. But when a definition that began life as a working hypothesis starts obscuring the path for researchers and doctors trying to find out what is wrong with a specific patient, one does need to ask whether further work might be needed on that working hypothesis.

Findings and Publication

One conclusion that emerged from our autism work was no surprise. Autism is a massive problem, and if you're going to have any chance of cracking it, you're going to have to mount a concerted effort with a team of experts from lots of different fields—a multidisciplinary approach, in other words.

This in itself is a problem, and partially explains why experts haven't had much luck finding out what autism is, let alone treating it. Taking on a complex multidisciplinary problem on its own terms, rather than sticking within the comfort zone of what you already knew, is brutally hard mental effort. It takes time and effort to identify and learn the relevant concepts, terminology, and basic facts from the disciplines involved. You don't need to learn everything from each discipline, just the relevant parts—but identifying which parts are relevant is not always easy, and when you have identified them, you need to understand not just those parts in isolation, but also the concepts and frameworks that anchor those parts into the bedrock of observed facts within that discipline.

That's one reason most cross-disciplinary research involves cooperation between teams that remain solidly based in their home discipline. A similar problem affects armies such as Hannibal's, where you're trying to handle groups with different languages and different ways of working. Hannibal didn't try to meld his Celtic cavalry and his Numidian cavalry into a single integrated unit—that just wasn't feasible under the circumstances. There was the Celtic cavalry, and there was the Numidian cavalry; the two were separate units. That makes sense if your options are limited, but it can go horribly wrong.

That's what happened to the German forces at Stalingrad in November 1942. Their defensive lines consisted of sections held by German troops interspersed with sections held by troops from other countries—a Romanian section, a Hungarian section, and an Italian section. The Red Army was well aware of the problems the Germans faced in trying to maintain tight coordination at the joins between sections, so when the Red Army launched its massive counterattack, it went straight for those joins and smashed right through.

There's a big-picture consequence. Most cross-disciplinary research problems are tackled using the "separate units" approach. If you're a neurophysiologist working on autism, you're still very much a neurophysiologist, and you'll almost certainly be based in a department with other neurophysiologists, even if they're working in completely different areas of neurophysiology. You almost certainly won't be based in the same department as, say, the pediatricians with whom you're collaborating in your autism research; they'll also be based in a department with others from their own discipline.

That makes a lot of administrative sense: for instance, if you're a university-based researcher, you're expected to give lectures in your subject, and the easiest way for the university to arrange that is to have clearly separated departments, each running courses in its own clearly separated field. It makes life simpler when assessing promotion and the like, where different disciplines often have different indicators of professional achievement, so being able to compare like with like has obvious attractions. It makes a lot of short-term practical sense, but it means there are clearly established fault lines built deep into the organizational system, with significant implications for research that has to span those lines.

Truly integrated interdisciplinary research has an inbuilt learning cost for the researchers—several years of focused work before they're really on top of the relevant concepts from the other fields—and very few researchers are going to invest that time and effort unless there's a realistic chance of an outcome that will make it all worthwhile.

So cross-disciplinary gaps in understanding, and misunderstandings, are going to be the norm rather than the exception for a long time to come. Governments and research funding bodies will make gestures toward fostering interdisciplinary research, but the learning times are against them.

Having a solid toolkit, like Verifier, of methods and concepts and language would allow researchers to communicate more effectively with colleagues from other disciplines, and there's a lot that could be gained from detailed studies of likely glitches in cross-disciplinary collaborations. The simple concept of laddering downward till you hit bedrock has worked in every field we've tried it in, and that approach meshes neatly with the well-established field of measurement theory.

The autism study uncovered clear patterns of *confirmation bias* in autism research, where researchers focus on evidence that agrees with their own view, regardless of whether it might be an even better fit with very different views. That's what sealed the Romans' fate at Cannae: if they had taken a mental step back and asked whether there could be other explanations for what was happening, they might have managed to get out of the trap before it closed.

Classic experimental design is intended to ask questions that reveal which of the competing models are consistent with the experimental findings. A lot of published studies weren't doing this, instead operating in terms of whether the evidence was consistent with the researchers' preferred model, regardless of whether it was also consistent with other models.

Other problems we found included the following:

- We've already noted the gaps in Kanner's original data, where potentially useful information about birth, development, and skills wasn't systematically recorded.

- There is a significant absence of precision in how the children's behaviors are described. Autism case studies contain a disproportionate number of distinctly non-scientific adjectives like *bizarre, odd,* or *strange.* You get the impression that the practitioner is aware that the behavior is abnormal but can't quite put a finger on why. An obvious solution would be to import some ready-made methods for fine-grained description and analysis of behaviors, borrowing from fields where this approach is routine. There are several fields that fit the bill, including ergonomics, psychology, and ethology. Finer-grained data would make it a lot easier for researchers to start answering questions such as whether some behaviors can be traced to a specific genetic cause rather than some vaguely specified "genes that cause increased predisposition toward autism."

- Several implicit, probably unconscious, assumptions in the early papers by Kanner and Asperger have been accepted without question by subsequent researchers. For instance, there is Kanner's dismissal of auditory problems as a significant factor and his framing of the symptoms in terms of "affect." That issue isn't unique to autism research; the initial framing of a research question often casts a long shadow on subsequent decades, regardless of the field.

- Use of poor categorization is widespread. Researchers tend to take a top-down approach, starting with concepts such as "social communication" rather than working bottom-up from specific, observable behaviors and symptoms. That has led to a categorization that fuzzes out into subjective terms and doesn't have a simple, unambiguous place for each specific behavior a child might exhibit. Working bottom-up offers the prospect of making much more sense of what is happening, and could be done relatively simply using established tools from other fields. Categorization is a key tool in research.

Our discoveries on autism were accepted for publication in 2009 in the top academic journal for autism. It was the second peer-reviewed outcome from Verifier, this time in a medical field. It was, for us, a significant success. We felt we had produced some important insights that researchers working in the field of autism could run with.

On one level, we had found no new pieces in the puzzle of autism using Verifier; what we noticed had been there all along, in front of everyone's noses. On another level, we had found something new, namely a different way of configuring the known pieces of the puzzle. We had also found a prime example of how researchers were often unaware of highly relevant issues in their own fields, never mind other fields.

Our findings produced some classic examples of people in different communities not communicating. One possible explanation comes from what we know about experts. The seven to ten years it takes to become a real expert in a field means there's a huge learning curve. You need to learn the concepts and terminology, which takes years and is usually brutal and exhausting.

If you're a special needs worker in a school with a grim workload, and you have the choice between reading a newsletter from a local autism support group, or a magazine article about dyslexia written in language you're familiar with, or an uncompromising, heavy-duty research article where every sentence contains terms you've never met before, you are likely to go with the easy read. As a result, the dyslexia "community of practice," which consists of people working in school support, does not connect well with the "community of discourse" for the psychology of reading, which consists of heavyweight theoretician psychologists. Very different communities; very different language and concepts; very different implications. Little wonder there's comparatively little exchange between them.

Again and again what emerged from our work was how important categorization was to expertise; experts typically have richer, better-organized categorization than novices. It's hard to imagine how any field could develop without using categorization. How you define something, how you lump it together with similar phenomena, how you split it from different phenomena—these eventually determine how you think about it.

However, categorization is not simple, and here a nearly right model can have far-reaching implications.

Categorization and Medicine

Most people don't give much thought to how medical diagnosis works; there's usually no need for them to think about it. Medical researchers, though, have spent a long, long time working on diagnosis, and they're very good indeed at most parts of it. At other parts, though, the classical medical approach clearly hits problems. The breakthrough successes of modern medicine came with germ theory and bacteriology, where you could put a medical sample through an objective test and say, yes, this patient has cholera or, no, this patient doesn't have cholera. It was a major development, and it changed medicine forever.

That model worked well not only for bacterial diseases but also for viral ones. The result was a model that was a good fit with reality, worked well with systematic large-scale tests of possible treatments, and generally had a lot going for it. The obvious temptation was to apply that model to *all* medical problems. And that's where it started to bog down. Here's why.

If you're dealing with bacteria, nature has already divided up the problem into pretty neat categories for you. There are different species of bacteria, just as there are different species of fish or bird, usually with fairly clear dividing lines. Ask a keen birdwatcher what species they saw last weekend, and they'll tell you about it in terms of this species of eagle and that species of songbird, but they won't tell you that "most of them were predominantly eagle with a strong strain of owl and a hint of robin."

Note the phrasing: "*pretty* neat" and "*fairly* clear." When you start to look in more detail, reality becomes a bit more fuzzy. Try asking your birdwatcher friend how many species of gull there are. There's debate about which species are truly separate and which are variants of the same underlying species—and that's with birds, which are comparatively easy to model. It's much worse in microbiology, where you are trying to distinguish between types of bacteria that can swap DNA with each other.

The picture gets more complicated with viruses, which have a nasty habit of coming in different strains, but it's largely true there, too: to a fair extent, viruses are already separated into categories by nature. The same applies to bits of your body breaking or wearing out; we all have much the same body plan, though the details vary. If you find the idea of significant variance hard to believe, try looking at a diagram of how the aorta, the main blood vessel coming out of the heart, is plumbed into other blood vessels. Textbooks typically show one arrangement, but there are actually about a dozen fairly common variants of that plumbing, and that's for one of the biggest and most important blood vessels. On the whole, though, the mechanics of the body are pretty similar from one body to another, and the genetics behind the mechanics are based on the same DNA architecture, so there are fairly clear distinct categories like "a blood vessel rupturing" or "an exit tube from the kidney getting blocked."

It all starts to fall apart the moment you try tackling "mental illnesses." The fact that I'm putting the term in quotes is a give-away that even the name of this field is a minefield. Yes, there are some conditions where there's a very clear mechanical or chemical problem with the hardware of the brain—a tumor here, an enlarged blood vessel there, an abnormally low serotonin level somewhere else. But a lot of the time, you're dealing with behaviors that don't map neatly onto anything tangible. And you're usually dealing with a bunch of behaviors, not just a single one. So it's hard to know where to start.

If an underlying genetic disorder is causing some or all symptoms, you can focus on that. This concept is well established and does correspond to reality; Down syndrome is one common example, and Williams syndrome is another, both linked to clearly identifiable genetic markers.

Beyond that, you are forced to deal with a vast variety of symptoms, like the ones Sue found in her analysis of autism. In such cases, it arguably makes more sense to consider each symptom as a condition in its own right, rather than trying to cure some higher-level "syndrome" that might be the product of some diagnostician's imagination. That's how developers debug software, which is a strong analogy if you view psychological problems as being akin to glitches with the software of the brain. You might take into account that some types of software error have predictable side effects and cause other software bugs in a ripple effect, but you wouldn't start trying to identify syndromes among those bugs.

Here's one way to tackle this issue of co-occurring features. It's an idea that I developed with John Chalmers, an exobiologist who has worked with NASA and who researches possible alien life-forms. A long-standing problem for exobiologists is how to recognize life-forms, including ones that might be very alien. A lot of thought has gone into this debate over the decades, with rival definitions being kicked back and forth.

John and I used one of the old, well-worn tools from the Verifier toolkit: take the obvious and reverse it. Don't look for *a* definition of *life* in the singular, but instead think about the opposite. When you phrase it that way, there's an elegantly obvious answer. You plot each of the most widely accepted definitions of *life* as a series of permutations. It would look something like this, where each column represents a different type of "life" or "sort-of-life" (figure 32).

When you do this, something jumps out at you immediately. This framework gives you neat, ready-made slots for some things that cause problems for single one-size-fits-all definitions of *life,* such as viruses and prions. Viruses and prions are entities that caused a lot of trouble for the medical researchers who first encountered them. In both cases, it took a long time—decades for viruses, years for prions—before they were identified and understood, partly because they didn't fit into the mental pigeonholes of "living" and "nonliving" that researchers were using. Viruses and prions are killers—think of AIDS and BSE, generally known as "mad cow disease."

	Category						
Feature	A	B	C	D	E	F	G
Metabolism	•	•	•		•		
Reproduction	•	•		•		•	
Responds to stimuli	•		•	•			•

FIGURE 32

Suppose that a Victorian researcher had doodled a diagram like the one you've just been looking at. Medical researchers would have been mentally prepared for discovering viruses and prions, potentially saving years of false leads. They would have said something to the effect of, "Aha, at last we've met the thing that fills that slot!" and immediately been able to get on with tackling it.

There's a compelling precedent for this approach. That's exactly what happened in chemistry, where the Russian chemist Dmitri Mendeleev produced a table of the elements, the periodic table, which chemists used in just that way, predicting the properties of as-yet-undiscovered elements and then searching them out. In that table, the elements are arranged according to their atomic number. Other scientists had tried to do the same thing, but Mendeleev's genius was in finding a way that handled the known elements as well as systematically predicting elements that had not yet been discovered.

An added irony for my team is that he did it using a form of card sorts. So, are there new categories of life or sort-of-life waiting to be discovered? We don't know. Like anyone who was enthralled by science fiction as a kid, I'd love to know. Maybe some of John's colleagues at NASA will find the answer. . . .

Joining the Strands

AND SO WE ARE LEFT WITH A RECURRENT PARADOX in science: experts are human, and humans make mistakes, and that's both good and bad. On the one hand, you want experts to make mistakes, because each new mistake brings them closer to solving a previously unsolvable problem. Experts, remember, make more mistakes than novices, because they use the results from mistakes to close in on the truth. On the other hand, you hope they *don't* make mistakes, because we all desperately need what the world's smartest people have to offer.

But what happens when they don't realize they've made a mistake? What happens when they don't notice that they are trapped in a logical maze of their own making? What happens when they keep banging their heads against the same problem for decades with no results? When that happens, everyone loses. If the problem is Alzheimer's, some people are slowly robbed of their minds. If the problem is autism, some people spend their lives distant from those who love them. If the problem is faulty bridges or misdiagnoses or malnutrition, people die.

We got lucky with Verifier, although that story didn't turn out the way we expected. We took on two significant problems in very different areas and spotted significant flaws in the received wisdom of each. In both cases, our critical reassessments of the field passed peer review and were published in significant journals in the relevant fields. At one level, that's two successes, both on shoestring budgets and in our spare time.

At another level, though, those stories still aren't over. The lessons we learned from both studies are shaping the next generation of Verifier work. In this chapter, I'll talk about the challenges Verifier faces going forward, but before I do, I might mention how we wrapped up the Voynich work and what lessons we gleaned from it.

Voynich: Carbon 14 and the "No-Kelley" Option

The Schinner paper and the Austrian team's failure to find signs of erasure or correction in the manuscript might have looked like the end of the story. From two completely different routes, these findings provided strong support for the manuscript being a hoax that contained nothing but meaningless gibberish, probably produced by Kelley. There was, however, another twist in the tale.

In 2010, a University of Arizona team conducted carbon 14 dating tests on some fragments of pages snipped from the Voynich Manuscript. Trace amounts of carbon 14, a naturally occurring isotope of carbon, linger in the bodies of plants and animals after death and decrease predictably over time. This predictable rate of decay allows scientists to work out the age of organic specimens, such as the pages of the Voynich Manuscript.

The Arizona team dated the manuscript's vellum pages to most likely between 1404 and 1438, the early end of the fifteenth century, which is over a century before the con man Edward Kelley was born. Though I liked Kelley for the hoaxer, he's not an essential part of the hoax hypothesis. But the new data gave me a chance to rethink my overall work and see how it fit with the new facts.

The date took everyone by surprise when it reached Voynich bloggers in 2009 and then hit the mainstream media more than a year later. The Arizona finding answers only one question, namely when the vellum was produced. That's a good start; it lets us eliminate any theories that claimed a much earlier date for the manuscript (though a few people have argued that the manuscript is actually a fifteenth-century copy of an older document). However, there weren't many claims that the manuscript was significantly earlier; most researchers believed it dated from a significantly *later* period, maybe the 1460s or early 1500s or the 1580s. The trouble is that the carbon date doesn't help much with those.

The carbon date only tells us when the vellum was made. More precisely, it tells us when the animal whose skin ended up as the book's paper was slaughtered. It doesn't tell us when the manuscript was written on that vellum. One of the team members studying the manuscript apparently said that the ink was written onto "fresh" vellum, but that description is unfortunately ambiguous. On one hand, it can mean "brand-new." But, in the context of parchment and vellum, it can *also* have a completely different meaning, namely "not written on previously and then scraped clean." Many old manuscripts were written on previously used parchment. Scribes scraped and washed the pages of old books clean and "recycled" the pages, presumably because it was cheaper than buying new vellum. But that hadn't happened with the Voynich Manuscript.

The key question is when the ink went onto the vellum, and that's a hard question to answer. You can't decide based on appearance. The trouble with vellum and parchment is, ironically, that they can stay in as-new condition for centuries, unlike most paper. The U.S. Declaration of Independence, Constitution, and Bill of Rights were all written on parchment. The British parliament still uses parchment as the standard material for recording laws, precisely because it is so durable. Documents from a thousand years ago may still look new, particularly if they've been kept in a good environment and haven't been heavily handled. If a modern-day calligrapher were given a sheet of parchment and asked to guess how old it was based on appearance, he would be lucky to get even the century right. It's perfectly possible that the manuscript could have been written on vellum that was already old, perhaps a century or more old.

That raises the question of how likely it is that someone fifty or a hundred years later would have enough already-old vellum to produce the manuscript. It's more likely than you might think. Voynich researcher Rich SantaColoma investigated this question and found evidence that old vellum was easily available both in antiquity and more recently.

A paper published in 2002 describes what happened when the researchers tried carbon-dating late medieval documents whose date was known from the headers in the documents themselves—"written this 15th day of March, 1482" or whatever. For each document, the authors show what are known as the one-sigma and the two-sigma values. That translates roughly into "there's a very high likelihood it's within these dates" and "there's a high likelihood that it's within this *broader* set of dates." They found that the document with a known date of writing in 1457 fell within their one-sigma date of 1444–1488; the same with their document from 1446, which fell neatly within their one-sigma range of 1438–1466. But three other documents didn't fare so well; they all fell into the broader range of dates. One, with a real date of 1434, fell *outside* the two-sigma range of 1437–1620; another, from 1434, was only just within the two-sigma range of 1321–1437; the third, from 1495, fell just within the two-sigma range of 1491–1645. In other words, even with a wider carbon date estimate spanning over a century, the real dates fell only just within the estimated range. Carbon dating is like a lot of other techniques: it's very powerful, but it's easily misinterpreted as a magic bullet. You need to understand it properly if you're going to use it.

Rich also found a website that, a few years ago, was offering for sale sixty unused pages of sixteenth-century parchment. That's enough to produce a quarter of the Voynich Manuscript, and it's an age gap—about five hundred years—far wider than anyone is proposing between the date of the Voynich Manuscript's vellum and the date when the ink went on. When I asked a couple of master calligraphers how common it was to find old parchment, they treated it as completely ordinary: they told me that they had parchment which had been in their files for more than twenty years, waiting for the right project to come along. So a substantial gap between vellum date and manuscript creation date is plausible.

The ink on the manuscript isn't much use for dating. It's oak gall ink, which has been in production using a more or less identical recipe for about a thousand years. I have a couple of bottles of the stuff in my own calligraphy kit, and they would probably be chemically indistinguishable from the ink in the manuscript, even though they were made only a couple of years ago. The Voynich pigments are consistent with a date in the early 1400s, but they're also consistent with a date well into the 1500s. In principle, it would be possible to carbon-date the ink on the Voynich manuscript, but that would take a serious amount of ink, more than the Beinecke Library would even think about allowing. Which leaves us where we were before the carbon-dating: trying to work out when the manuscript was created via other indicators in the manuscript itself.

One place to start is the lettering. The Voynich community agrees that the more flamboyant letters in the manuscript are inspired by a distinctive form of writing known as the humanist hand, which flourished during the mid-fifteenth century. Handwriting has fashions, just like clothing, and some are distinctive and short-lived, like the humanist hand. The general consensus in the Voynich community is that the writing style in the manuscript suggests a date between about 1450 and 1470. That makes a date much before 1450 implausible, but it does not prove that the ink went onto vellum that was already thirty years old or more, which is what we'd expect if the vellum dates to the middle of the carbon-dating range, around 1420.

Carbon dating tends to be misinterpreted by nonexperts. What you get with carbon dating is a probability statement. The date that comes back from the lab says, in effect, "There's an x percent likelihood that the artifact was produced between date 1 and date 2. There's a smaller, but still real, likelihood that the artifact was produced between this older date and this younger date, and an even smaller likelihood that it was produced between this very early date and this very late date."

When the University of Arizona announced a date that was generally taken to mean "between 1404 and 1438," that actually just meant that there was a good likelihood that the vellum dated from somewhere in that range. It also meant that a date of, say, 1450 was possible but less likely; you can calculate the precise likelihood from the date range and confidence figures in the university's original announcement (in this case, 95 percent, which is pretty tight).

Thus it's *possible* that the ink went onto brand-new vellum in 1450, but that's pushing the odds a bit. It means taking a low-likelihood value for the carbon date and an early date for the handwriting style. That could have happened, but it looks more likely that the ink went onto already old vellum. Once you accept that the vellum was probably already old, the next question is *how* old, and the cutoff is largely a matter of opinion. Could someone fifty or a hundred or a hundred and fifty years later have found an old book with only its first few pages used and cannibalized the rest to produce the Voynich Manuscript? The answer to that is a clear yes; it's possible. There are old books out there with only the first few pages used, and there are record books and chronicles that were in continuous use, sometimes at the rate of just a page or two a year, for centuries, so that scenario is far from impossible. It has actually happened at least once: Rich SantaColoma found a handwritten nineteenth-century book, the Chittenden Manuscript, that was written on centuries-old vellum by an American antiquarian.

Although dating on the basis of hairstyles and clothing and other imagery in the images of the manuscript is less solidly grounded than dating by the handwriting, there are grounds for suspecting a date well after 1430 from an assortment of directions. One famous example in Voynich circles is the so-called sunflower illustration (figure 33). It's a flower that looks a lot like a sunflower, with roots that could be interpreted as a potato's tubers. Sunflowers and potatoes weren't known in Europe until after 1492. No one knows whether that is coincidence. Arguments have been advanced and denied for decades, and there isn't much prospect that they'll be resolved any time soon.

The hoax hypothesis is not particularly affected by the new dating. Nobody will seriously claim, "Hey, the book dates from 1430, therefore it can't be a hoax, because everybody knows that you don't get hoaxes until the sixteenth century!" The carbon dating doesn't tell us whether the manuscript contains a code or a language or a hoax. It does make Kelley unlikely to be the hoaxer, but there were other hoaxers in the world long before he was born, some known, some unknown. Confidence men aren't confined to one period of history. One may have created the manuscript, but we simply can't tell from a carbon date for the vellum.

The particular mechanism that I suggested, the table and grille, was an excellent fit for a late date, since the Cardan grille

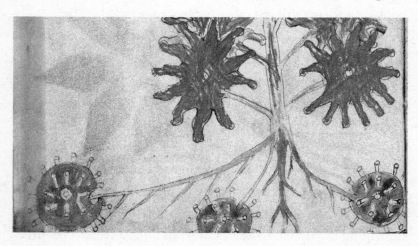

FIGURE 33

wasn't invented till 1550. If the manuscript was written earlier than that date, the grille is out. Was there some other effective tool that would have suggested itself to a wily hoaxer? There is indeed another old technology that involved moving a piece of card with holes in it across an underlying sheet of paper. It was the stencil, used to mass-produce playing cards. That method became widespread in northern Italy, the source of the vellum for the Voynich Manuscript, around—you guessed it—1430. So, ironically, a method *similar* to the table and grille hoaxing method is comfortably compatible with the unexpectedly early date for the vellum.

There are a few other tantalizing possibilities, if you go with the idea that the manuscript was created as a meaningless hoax using tables and grilles. One is that the playing card stencil model is a better fit than the simple grille I used, given some odd features on the first page of the manuscript. The earliest Voynich researchers noticed that the first page of the book is crammed full of features that are rare elsewhere in the manuscript. One particularly flamboyant character, the Currier *Y* (figure 34)—named after a U.S. Navy cryptographer, Captain Prescott Currier, who studied the manuscript in the 1970s— occurs several times on the first page but then occurs only once every few pages thereafter.

Such a rarity might make sense if it occurred within the same word each time. You could imagine someone repeatedly using a particular word containing a rare character in an introductory section while mentioning something important that will reappear later. But that's not what happens on the first page; instead, the mysterious Currier *Y* occurs in different words each time, at various points in the line and not just at the start, which would happen in an alphabetical index. That distribution

FIGURE 34

is hard to reconcile with an unknown language containing real meaning. It might make sense in terms of a code that was disrupting the letters in the words where it appeared, but nobody has been able to come up with a plausible code that would do that and still resist decipherment. It's a lot easier to explain it in terms of a hoaxer playing around with the characters and the technology.

I wondered whether this might be a weak spot from text produced before the hoaxer spotted the weaknesses, which might allow me to reconstruct a fragment of the underlying table and grille. I found that it was indeed possible to produce something that sort-of generated those words, but it appeared that a different grille was being used each time. That's plausible; if you're using the table and grille method, you'd use a lot of grilles sooner or later, and you might experiment at first using different grilles within the same page. You would soon discover that it's less hassle to stick with one grille for one pass across

the table. In addition, there is less risk of accidentally producing identical text fragments on different pages of your hoaxed
manuscript.

The occurrence of the Currier *Y* was a better match, though,
with a scenario where someone was using a grille modeled on a
playing card stencil, specifically one of the higher-value cards,
where the stencil would contain numerous holes. I had assumed
that the hoaxer would use a card with the minimum number of
required slots—one each for prefix, root, and suffix. But if they
instead used a card with two or three slots each for prefix, root,
and suffix, then arbitrarily chose one prefix slot, one root slot,
and one suffix slot, that would produce the apparently incon-
sistent outputs that neatly minded cryptographers hate because
it doesn't fit into their worldview, but that a hoaxer trying to
disrupt regular patterns would love. That doesn't narrow down
the date when the book was written, though, since the playing
card stencil could be used any time after its invention, even a
century and a half later.

Thus it's perfectly feasible that the manuscript could have been
produced in the 1430s using a table and grille method, just not
Cardano's. However, tantalizing clues point in the opposite direction, toward a much later date. One feature of the manuscript
that suggests a date well after 1430 involves the "names" on the
plant pages. Each plant page begins with a unique word, apparently the name of the plant. That feature has been a magnet for
cryptographers, who have spent decades trying to make sense
of what's going on there, without success. The most widely accepted theory is that the words are some sort of signal to would-
be decipherers, telling them something like "now shift to cipher
4." That already suggests a comparatively late date; it's a sophisticated approach, hard to reconcile with an early date.

However, that feature is also exactly the sort of bone that a hoaxer with a knowledge of cryptography would throw to would-be code breakers. Proper nouns—names of people and places and the like—are the bane of a code maker's life. They are something that code breakers home in on like wasps to a honeypot, because they're the classic weak point in a cipher. It's the same with unknown languages. That's how the nineteenth-century French scholar Jean-François Champollion cracked the hieroglyphs inscribed on the enigmatic Rosetta stone: he guessed that some words, enclosed in a symbol known as a cartouche, were probably royal names. Since he knew the royal names, and they were a manageably small set, he was able to crack that part of the hieroglyphics and make a way into the rest of the ancient Egyptian. So, if you put something that looks like proper nouns into your meaningless hoax, that will make code breakers more willing to believe that they're dealing with a real cipher text, and will also focus their attention on that blind alley instead of on places where they might figure out what was really going on. How long would it take to generate a hundred-plus unique names with linguistic features similar to, but subtly different from, the text in the rest of the manuscript? Answer: about a couple of hours. It's that easy. I know because that's how long it took me when I tried.

If that really is what's going on with those "names," it implies that the hoaxer was familiar with reasonably sophisticated code breaking. That's hard to reconcile with a date of around 1430, when codes weren't complex, but is a much better fit with a later date, when cryptography was more sophisticated. It probably wouldn't be significantly earlier than about 1470, and could fit with almost any date after that.

Some have claimed that clothing and hairstyle features in the manuscript would fit best with a date in the early 1500s. There's someone who might just fit the bill for hoaxing the manuscript around then: Girolamo Cardano, inventor of the Cardan grille, who was born in Pavia, in northern Italy, in 1501. A brilliant, complex, difficult individual, he was an expert on codes, and there's an intriguing possibility that he began as a young man by hoaxing the manuscript with an early form of the table and grille, and then realized in middle age that the same grille could be used for coding genuine cipher text. Cardano knew John Dee, Queen Elizabeth I's consultant, so it's possible that the Voynich Manuscript originally came from Cardano—either made by him or perhaps inherited from one of his scholarly ancestors—and then found its way into Rudolph's court via Cardano's contact, John Dee. Maybe. It's conceivable, but it's speculation, without any solid evidence.

The last set of features in the text is also speculative, but it makes sense of some otherwise puzzling loose ends. The most distinctive features of Voynichese are its unusual script, its complete absence of word order, and its distinctive word structure. Each unusual feature is the mirror image of a distinctive feature of Enochian, the alleged language of the angels invented by Edward Kelley.

By "mirror image" I mean this. Enochian has three highly distinctive features. The first is that Enochian script, though beautiful, is a pig to write, especially in a hurry. The second is that the word order of Enochian maps directly onto the word order of Elizabethan English. The third is that there's no particular word structure within Enochian words, so an Enochian word can end up with an unpronounceable string of consonants clustered together. Kelley was a novice at creating languages—a brilliant one, but still a novice, with only one language under his belt.

One characteristic of novices is that they overcorrect when they make a mistake—think of what it's like being behind a student driver as he veers from one side of the road to the other. So, if Kelley went on to produce a second language after Enochian, it's a fair bet he would overcorrect those three annoying features of Enochian and produce a language with distinctive features that went in the opposite direction. He would create a language that was *easy to write,* that *didn't have the same word order as a real language,* and that had a *nice, regular word structure,* like prefix/root/suffix. If those features sound familiar, it's because they are; they're three of the most strikingly distinctive features of Voynichese.

That speculation may be significant, or it may be just coincidence, the same way that it may well just be coincidence that the manuscript appears at Rudolph's court at the same time as Dee and Kelley just happen to make the eight-hundred-mile trek from London to Prague, with no hint of the book's existence at any point before then.

There's one more closing coincidence, again from the diligent fieldwork of Rich SantaColoma. Kelley was interrogated in 1592 by Rudolph's men, and records describe what happened. One question was, "What was the signification of the secret characters in Kelley's note-book?" It appears they were unable to extract an answer from Kelley. Rich is hunting down the original documents, in the hope that they'll make sense of that tantalizing remark. Were those "secret characters" more of Kelley's codes? Or was he reluctant to admit what they were because actually they were his production notes for the Voynich Manuscript? Maybe, just maybe, Rich will track down the original records of that interrogation in a musty archive, and maybe they'll give us a final answer. It's a long shot, but well worth a try.

When Voynich discovered the manuscript, it seemed too complex to be a hoax, too bizarre to be an undiscovered language, and too uncrackable to be a code. Now we at least know that a hoax is feasible, so there's one simple, unmysterious explanation for what the manuscript could be. For what it *could* be. As for conclusive, knock-down proof of what it really *is,* though, that's another story. There's a logical asymmetry that's a real irritation when you're dealing with a possible hoax: you can prove a code by deciphering it, you can prove a language by translating it, but you cannot prove a hoax. Proving an absence of meaning is a lot harder than proving its presence; there's always scope for someone to say that you just haven't looked in the right place. And in this case, a lot of people have said just that.

It looks as if the Voynich story will continue to fascinate scholars and hobbyists and the public for a long time to come. It bewitched Rudolf II, bedeviled Wilfrid Voynich, taunted the World War II code breakers, and now vexes modern scholars. It will almost certainly be intriguing people decades from now. It's a fascinating problem and an elegant one, but I'm not planning to put much more time or effort into it. There are plenty of other challenges that have bigger implications for the world, and the Voynich case study had given us some good insights into what we needed to do next with our expert-checking method.

Verifier and the Problems That Science Tackles

Looking back on the Voynich and autism case studies, it was clear that Jo Hyde and I had been half right and half wrong in our guesses about how the rollout of Verifier would go. That had deep implications for where we might, or should, go next.

We had expected maybe one case in ten to produce a significant result, as measured by getting a proposed new insight through peer review, if we were lucky. Instead, Sue and I produced two significant results, one for each case we investigated. That's a good hit rate. Getting published in a leading journal in a field that's new to you shows that you're doing something right.

But that didn't mean our findings would produce an instant, obvious change in those areas. Research doesn't work like that. If you look at some of the most important discoveries in the last century, you will find that they were usually controversial for years, and quite often rejected for decades before finally being accepted. Sometimes that's because of professional disagreements—scientists are human, too. Sometimes it's because of low-level infrastructure. In medical research, for example, any new pharmaceutical drug has to go through years of tests and trials before it can go on the market. There are also issues such as replication, where other researchers try your new idea to see whether it really works. That takes time. Changing key assumptions in a research field such as autism isn't a swift, dramatic act like the swerves in a Hollywood car chase. It's more like getting an ocean supertanker to change course, where the sheer size of the beast means that change is going to take time.

More interestingly, Jo and I had also been wrong about how Verifier would work. We had assumed that it would be a very slow, painful process, the equivalent of surveyors slowly mapping out new territory yard by yard, in meticulous detail. Instead, it had been more like explorers traveling hard and light across brutal unknown terrain, machete in one hand and journal in pocket, coming back exhausted with a sketch map that showed a passable route through the jungle and mountains and might lead to the plains beyond.

Ironically, we had ended up using the same behaviors that allow experts to work much faster and better than novices but also produce the predictable errors that Verifier was designed to catch. We did that because some of the things that turned out to be critically important couldn't be done any other way.

The autism study is a good illustration. I'll go through the stages and explain what we learned about each step.

1. *Decide if a problem really exists.* The first stage of Verifier can't be done using explicit, step-by-step logic and big, formal diagrams. First, you're looking for indications that the apparent problem might just be another moral panic—the equivalent of the fear that soccer matches have reached a new high in violence—or that the presenting symptoms are very different from the underlying cause, or that there's some social framing (looking at the world from a particular direction) going on. You learn to recognize the patterns with time.

For instance, if you see claims that something is happening now at unprecedented levels (like those claims about soccer violence, or the claims that there's an epidemic of autism), you immediately wonder whether it's a moral panic or a conspiracy theory, and you look at the type of evidence and reasoning and knowledge being used, as well as the expertise of the people making the claims. As a rough rule of thumb, if the claims are coming from a large number of researchers in different institutions, all using well-established methods with a significant number of peer-reviewed papers behind them, then it's time to start taking them seriously. They might turn out not to be correct in the end, but you can't simply dismiss them straight off. On the other hand, if the claims are coming out of politics or from concerned amateurs or from people whose expertise is in an unrelated field, then it's time to get suspicious. They might happen to be right almost by accident, but that doesn't happen very often.

This stage is invaluable—you're checking for whether there actually is a problem to tackle in the first place—but it doesn't need formal diagrams and notations. Instead, it uses a lot of pattern matching.

2. Try to determine if the problem has already been solved by experts in a different discipline. The second stage requires a different type of pattern matching—because you're trying to spot things that aren't there. Once you've decided that there is a real problem to investigate, you need to check whether it has already been solved, and whether there are any significant gaps in the types of expertise that have previously been brought to bear on the problem.

If you're dealing with a problem in business or industry, you quite often encounter the "already-solved" issue. A lot of problems corporations and manufacturers grapple with have already been solved somewhere else, though usually not in a place where anyone would think of looking. It's often quicker to solve the problem from scratch than to look for a ready-made solution somewhere else, especially when you don't even know if there is a ready-made solution to be found. One case I bumped into involved an industrial manufacturer who wanted to know how to prevent teapots from dribbling tea when people poured from them; I happened to know that the underlying cause of this problem was well understood in the field of medieval pottery research. Not exactly the obvious place to look.

How can you work out where to start looking, when the list is potentially infinite? One way is to look at the deep structure of the problem, and then think about other areas that are likely to have that same deep structure. Again, you're using pattern matching.

3. Determine if there are any expertise gaps. Spotting expertise gaps is sometimes easy. Sometimes when you examine the deep structure of what they're trying to do, you recognize it and do a double-take: "They're trying to do that without using *X*." When it comes to recognizing expertise gaps, the usual suspects are two areas that human beings often have trouble with: categorization (choosing the right mental boxes to put concepts into) and representation (making models or diagrams to help people better understand concepts). Another thing that sets off alarm bells for us is whenever someone dismisses a possible explanation as a "dead end." Do they have expertise in that area? If not, they might well be mistaken, and you need to haul in an expert from that area to examine the claim.

You might start using some formal representations here, to show what methods and areas of expertise are currently being used in the field, but there's still a lot of pattern matching in this stage.

4. Identify the key themes and chains of evidence that people are using. In this step, you are focusing on the *key* themes and evidence chains—not the most widely used ones or the ones you think are most likely to be right—but the key, deep-structure ones. That's very different from the usual method experts use to survey the current knowledge in particular field: literature review, and the Systematic Literature Review, a highly codified type of literature review that interprets the word *systematically* in a very specific way. With Verifier, you're constructing a mental or physical network of the key cause-and-effect theories proposed by researchers, regardless of whether you agree with them or not, ignoring for the moment how much evidence there is for and against each link in that network—that part comes later.

Determining cause and effect is the chief job of science. But there's a big difference between finding what causes an effect and merely finding a *link* or *correlation*. These last two are mere suppositions. They're the starting point for research, the spark that perhaps prompts you to design an experiment or study.

Here's an example. Once an optician showed us a video of a Parkinson's patient with the typical Parkinson's tremors. The patient put on glasses with blue lenses, and within seconds the tremors stopped. That raised immediate questions about what was going on. The optician didn't know the cause, but he was sure there was a real effect. Could it be a guide toward a cure or new treatment?

If you tackle this mystery by doing a search of the medical literature, you find that there are respectable peer-reviewed papers about the mechanisms by which the human body's internal clock calibrates itself. These papers demonstrate that light is more blue-tinted in the morning, and more red-tinted in the evening. That raises the issue of how the body could use this effect to calibrate its internal clock. Other perfectly respectable papers tell you that the mechanism involves a pigment called melanopsin found in photosensitive retinal ganglion cells. Another batch of papers suggest that melanopsin suppresses melatonin production possibly via an enzyme involved in changing serotonin into melatonin. Melatonin suppresses dopamine production, and reduced dopamine is what causes tremors in most Parkinson's patients.

None of what you have just read *proves* that blue lenses will cure tremors in Parkinson's patients. Nor do those papers prove that it's a *likely* explanation. All they show is that there's a *possible* chain of causality. This search of the key themes says nothing at all about the strength of the logical links you joined to reach this conclusion, nor does it say anything about how big the effects are at each stage. It might be that the cumulative effect is actually miniscule; that's why this is just a first stage. Sooner or later, you need to answer all those follow-up questions and more, before you could assess how strong a candidate that chain of reasoning is as an explanation. Which is where the classic scientific method comes in; it was designed to answer questions like that, and it's good at it.

Tracing links like this may look like an obvious thing to do, but it isn't obvious, and it isn't easy, because it has a habit of taking you into different disciplines every couple of links in the chain. That means you have to figure out the deep structures in the reasoning of a discipline whose language and concepts may be unfamiliar to you, with the attendant risk that you'll get it grossly wrong, unless you're willing to invest time and effort getting yourself up to speed in that discipline. That's the downside. The upside is that you've got a chance of spotting connections that have been missed before. In the case we just looked at, the people working on Parkinson's aren't usually opticians, and vice versa, so they wouldn't be likely to find out about that particular set of connections unless they made a significant effort to look for them.

Ideally, someone using Verifier would trace only the concepts that are relevant to your problem; she doesn't have to take on the whole of each discipline that crops up, just the relevant parts. She has to unpack the explanations of the relevant concepts, and the explanations of the explanations, and so on until she hits bedrock. It's not an endless search, but that doesn't mean it's a gentle impressionistic stroll through the park either; researchers doing this need to be prepared for some hard learning curves.

This is a stage where formal notations and diagrams start to come into their own. You want to show systematically and clearly the chains of evidence and of cause and effect that are being used, and you want to show them in a way that lets you spot weak links and gaps as easily as possible. There has already been a lot of good work on this, much of it in the field of argumentation, but there's still a lot more that needs to be done.

5. Carefully study the concepts and representations used by experts in the field. "Concepts" can take such forms as game theory or clades or molecular structure—different fields use different conceptual building blocks. "Representations" might include hierarchy charts or pictures or equations or maps—again, different tools in different fields. And there's a reason why we say we are "mentally" mapping this all out. It's because you're trying to make sense of the big picture, while keeping it all in your head.

6. Deliberately avoid writing anything down or using formal diagrams too early. As soon as you start writing something down, it crystalizes in that shape, and you'll then find it that much harder to reconfigure the big picture into a different shape. Once we humans get a fixed image or notion in our heads about how the world works, it's hard to shake it loose. The neurophysiology for that is probably the same as the well-understood neurophysiology for the "tip of the tongue" effect, where the harder you try to remember a name, the more you get stuck in the rut of repeatedly remembering a *different* name that you know to be wrong. Nonetheless, you'll need to start using those formal approaches at some point; otherwise you'll just go on reading forever without getting anything done. It's a difficult balancing act.

7. Deliberately avoid actively looking for errors in the material you're reading; concentrate on getting a feel for the big picture. That sounds paradoxical, but there's a reason for it. If you start looking systematically for errors from the start, you'll grind to a halt immediately because there are so many apparent errors. One of my students, Boyd Duffee, wrote some software so I could annotate the errors and gaps in reasoning and the unsupported assumptions in a piece of text. That was an eye-opener. There were gaps in reasoning and unsupported assumptions all over the place in all the scientific papers I looked at. I couldn't see the forest of significant errors because there were so many trees of minor apparent errors.

It wasn't because the authors were scatterbrained. It was because of the low-level realities of how research papers and white papers are written. When you write professionally, you have to deal with word counts. You have to cram your message into a very restricted number of words specified by the publisher or your client, so you have to make some hard editing decisions. You omit minor detail. You simplify nonessential parts. You don't bother unpacking bits of reasoning that everyone in your discipline will understand. The result is that your text appears to be riddled with gaps in reasoning and unsupported assumptions. That means anyone trying to spot problems with reasoning and evidence has to make constant judgment calls about what's a significant problem and what's just an example of professional shorthand.

8. *Now home in on the things that feel wrong.* By this stage in the process, you'll probably have found at least one thing that rings your mental alarm bells. With my Voynich case study, there were two alarm bells. One was the widespread assumption that complex outputs required complex creation processes. The other was the absence of expertise about hoaxing. With Sue's autism study, one alarm bell was the lack of systematic

unpacking of definitions. This stage is where the balance starts shifting from pattern matching to formal, explicit reasoning using diagrams and logic and all the other relevant tools.

9. Choose tools that will let you check whether you really have found an error. Use those tools. If you've identified a real, significant error, and you've chosen the right tool, the answer is likely to jump off the page at you. With my two alarm bells about the Voynich case study, I didn't need much formal verification. Just citing the example of fractals was enough to show the error in the assumption that complex outputs require complex creation processes—fractals contain amazing complexity but arise from a very simple process. Showing the lack of expertise in hoaxing was equally easy. For Sue, some quite simple visual representations showed where widely used definitions in the autism literature were actually very imprecise. That last stage turned out to be easier than we had expected. Getting there, though, had not been easy.

Lessons Learned

Using Verifier this way is as brutal as it looks. It's not brutal just because of the sheer volume of information, or because it can take you across several disciplines, each with its own steep learning curve. It's also brutal because it involves a lot of mentally joining dots and spotting gaps, which isn't easy. We expected that. One less obvious problem was that the pattern-matching element in Verifier had serious implications for how you could get teams to use Verifier.

If you've read the literature on *X,* and you've also read the literature on *Y,* then you can spot the similarities and differences between them through pattern matching. However, if you've only read the literature on *X*, and your teammate has only read the literature on *Y*, then you'll only be able to spot similarities and differences by comparing *X* and *Y* using formal, explicit representations—exactly the sort of slow, laborious approach that Jo and I dreaded. Pattern matching—scanning large chunks of text and comparing it quickly to your acquired mental knowledge on a subject—offered a much faster approach. But it came at the price of depending on individuals who would take on a huge amount of hard work in the critical early stages.

I had been lucky for the first part of the Voynich case study; the absence of experts on hoaxing was obvious once I looked for it. The second part, finding a possible mechanism for a hoax, was more difficult but could have been a lot worse. Sue's experience with the autism case study was a lot harder, largely because she had to take on some substantial literatures that were new to her.

One of the big lessons of both studies was realizing how important human-based pattern matching is to the entire process. We hadn't expected that. Attempts by previous researchers to help experts avoid bias depended heavily on sequential reasoning, the kind of thinking that computers can handily perform and that fits well with team-based approaches rather than individual work.

We deliberately used Verifier without software support in those case studies, so we had more flexibility. But doing Verifier in your head the whole time is horrible—it's an enormous mental load. Sue, for example, surveyed the literature of autism, dyslexia, and chronic fatigue syndrome in a burst of activity that took nearly three years. She got remarkable results, but going forward, it's impractical to expect other researchers to do the same. It was clear that we would have to automate part of the process. That raised the obvious question of just how to automate it. Fortunately, our experience with the case studies and their spin-offs had given us some solid, practical ideas about where to go next.

Software and Verifier

If you look at how new products arise, a pattern emerges.

Stage 1: People have a bright idea for a better way of doing something.

Stage 2: They build a big, cumbersome prototype.

Stage 3: They build a new version that's much smaller and neater.

That's what happened with automobiles and aircraft and most other products. The early prototypes look like something out of a cartoon, and there's no way you could sell them to the mass market. That doesn't matter. The key point is to demonstrate that the underlying concept actually works. The Wright brothers' aircraft showed that powered, heavier-than-air flight was feasible. After that, it was just a case of improving the technology.

That's a pattern Jo and I had experienced firsthand with the Search Visualizer software. It started off as an idea for a better way of finding relevant records when you're doing an online search. The early prototype produced printouts that were huge and cumbersome—so big that I had a board installed just below the ceiling of my office to provide hanging space for printouts that stretched from ceiling to floor. There was no way the average user would accept that. The key point, though, was that we weren't planning to sell that version. The prototype had shown that we were onto something significant. It let users do things they couldn't do with any other technology. Now, we just needed to build a new version that was small and neat and user friendly. And that was something we knew how to do. We soon had a version that could easily display huge documents on a single screen, and you could learn to use it in seconds. We had completed the cycle of idea, prototype, and market-ready version with the Search Visualizer. The next question was whether we could do that for Verifier.

The obvious way to tackle it was the way Jo and I had first envisaged—a huge, sprawling piece of software that mapped out every assumption and piece of evidence and step of reasoning that related to the problem. There are some obvious drawbacks to that approach, like the huge effort the user would need to put in, but those drawbacks don't mean it's wrong. Doing proper statistical analysis is also difficult and has a horrible learning curve, but that doesn't mean people should stop using statistics—on the contrary, good statistics are an essential part of science.

We considered that possibility, and then we used our old friend, the technique of taking the obvious and reversing it. The obvious approach was *a single piece of software* that was *big* and *designed for a few specialists.* If you reversed that, you got *multiple pieces of software* that were *small* and *designed for large numbers of nonspecialists.*

That multiple approach had a lot of attractions. We had deliberately done our first two Verifier case studies the hard way, by hand, precisely because we needed flexibility in how we tackled the case studies. Big software systems may force you to do things the way the software wants, or they may be so complex that you have to spend more time learning about the software than about the problem you're trying to solve. They're also horrible to produce, especially when you're trying to get the component parts to fit together—for instance, getting the part that handles knowledge elicitation to fit with the part that handles choice of visualization method. With the multiple software approach, we wouldn't have those problems. We could make each piece of software small and light and easy to use; "joining the dots" between separate pieces of software could be handled via human pattern-matching skills, which could do things that would be difficult or impossible for a computer using sequential processing.

There was another advantage as well. We had designed Verifier to *find errors* in work *that had already been done.* If you went down the route of multiple pieces of software that were small and could be used by nonspecialists, it could be used to *prevent errors* in work *that hadn't yet been done.* That would have a lot of attractions to people in a lot of fields.

That takes us back to a topic I had first encountered years before—the idea of software that would help experts and decision makers avoid mistakes (often known as Decision Support Systems, or DSSs). The DSS idea had been around for decades but hadn't gained much acceptance. However, we now have a different way of looking at the problem. Like traditional online search systems, the old Decision Support Systems were based on step-by-step processing, which was slow and cut across the grain of how experts normally work. With our framework, though, we can tackle the task via parallel processing and pattern matching, which is much faster and easier for people to use, and will probably also be able to do things the previous systems couldn't do, just as the Search Visualizer opened up new possibilities within online search.

If we tackle the problem this way, we could produce affordable tools that any decision maker or researcher or interested member of the public could use—tools that would illustrate graphically exactly what someone was trying to say, and would reduce the risk of misunderstanding or dodgy reasoning. In addition to the Search Visualizer, there could be a Categorization Visualizer and a Definition Visualizer and a Chain of Reasoning Visualizer—it wouldn't need many tools to cover the key points.

Here's an example. It involves something most people assume to be simple: ways of defining *male* and *female* (figure 35). The traditional view is that someone is either male or female, with no in-between categories—the black and white in definition 1. But, as we saw back in chapter 4 when we talked about androgyny theory, that's not the only possible definition. Another possible view is that some people are definitely male, some are definitely female, and some are on a sliding scale between the two, as in definition 2. In definition 3, everyone is on a sliding scale from very male to very female, with no distinct cut-off point in the middle. Another option, definition 4, is that some people are definitely male, some are definitely female, and others fall into one of several clearly defined categories with no sliding scale—for instance, people who are chromosomally XXY or people who have a specified endocrine condition. Other definitions are also possible—we've just shown a few.

The key thing about this type of visual representation is that it lets people specify unambiguously which definition they are using. That reduces the risk of misunderstanding from the outset. It also means that the person's assumptions are clearly stated in a way that makes it much easier to test the correctness of those assumptions.

Definition 1: Binary either/or

Definition 2: Male, in-between, female

Definition 3: Grayscale continuum

Definition 4: Mosaic—male, various clearly defined in-between categories, female

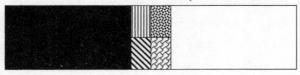

FIGURE 35

That's one example of how simple visual software can help clarify an argument. That raises an immediate question about just how many pieces of software would be needed, and how we would know what they should be. That's a question about requirements, and requirements engineering is familiar ground for me. It didn't take long to start getting solid traction on this problem.

The field where we're applying these methods is about as de-manding as anyone could ask for. It's about explaining forensic evidence to ordinary members of the public. If you can handle that challenge, then you should be able to handle pretty much anything. Here's why it's difficult, and what we're planning to do.

Forensics, Evidence, and Visualization

Explaining evidence to jurors is a big problem for forensics experts. The current method is based on probability, which is tricky. For instance, if you say "There is a 0.0167 percent chance that the suspect did not commit the crime," most people would have considerable difficulty understanding such a statement. The situation is even more difficult when the same evidence points in different directions with different strengths. For instance, if the crime was definitely committed by a left-handed person, and the suspect is left-handed, that evidence is 100 percent consistent with the suspect being guilty. But on its own, it is extremely weak evidence, since it's *also* consistent with millions of other left-handed people on the planet being the criminal. Sometimes a single source of evidence, such as a footprint, might contain several features that are consistent with the suspect being guilty, and others that are consistent with the suspect being innocent.

A lot of people have trouble grasping what those probability estimates mean. A common mistake is that nonexperts still expect a pair of cut-and-dried figures to give absolute limits, such as the earliest and latest possible times the death could have occurred, even when the expert is explicitly saying that there isn't a clear-cut answer.

We wanted to help the forensics people get around this problem. We knew there was a substantial body of research consistently showing that people are much better at handling probabilities when they are presented in other ways, particularly when they are presented as frequencies. This is now accepted as best practice in some fields, particularly some areas of medicine, but hasn't penetrated some other fields, such as forensics.

We realized that you could show the numbers visually in a way that's systematic and also easy for people to grasp using pattern matching, with no math involved. Here's how you can use that approach to show an estimate of time of death—a common problem in murder cases. The usual difficulty is that juries latch onto the times at each end of the estimate, just as people latch onto the dates at each end of a carbon 14 date, and lose sight of the likelihood issue.

The alternative is simple (figure 36). The darker the shading under a given time, the greater the likelihood that the death occurred at that time. With this representation, it's obvious that there's no clear cut-off point, but it's also obvious which times are reasonably likely and which times aren't. That makes it a lot easier for the jury to grasp. For the experts, there's another attraction. This representation is based on a systematic mapping between features of the probability estimates and features of the diagram itself. The darkness of any given point on the bar corresponds systematically to the statistical probability of death occurring at that point. This means the image does not

FIGURE 36

involve any risk of distortions due to artistic interpretation. It also means that the image can be automatically generated by software.

This concept has sparked a lot of excitement among police and forensics people who have seen it. It's just one of the visualization methods in the toolkit. Here's another example, based directly on an approach developed by Gerd Gigerenzer and his colleagues to help doctors explain medical probabilities to patients.

A common problem with forensic and medical tests is how to show true and false positives and negatives. If you use a picture that contains very large numbers of small squares—typically at least ten thousand—you can easily show how big a risk really is. In figure 37, the shadings show whether the squares are incidences of true positives, false positives, and so on.

This example shows just a hundred squares for simplicity; the real thing would typically have thousands or tens of thousands. I've used the following color coding for this example.

True positive = black.
The chance that the test has correctly shown a positive result.

False positive = checkerboard pattern.
The chance that the test has incorrectly shown a positive result; the correct result would have been a negative.

True negative = white.
The chance that the test has correctly shown a negative result.

False negative = wavy lines.
The chance that the test has incorrectly shown a negative result; the correct result would have been a positive.

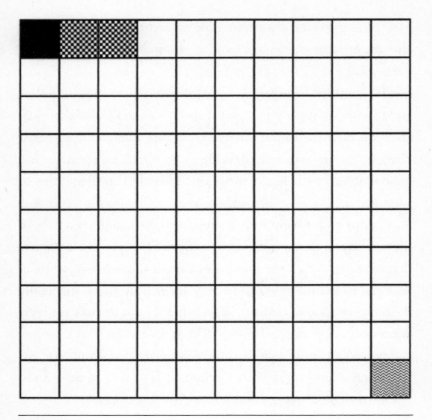

FIGURE 37

Forensics and medical experts often struggle to explain these odds to laypeople. If this example was a forensic test, for instance, and the test came out positive for the suspect having gunpowder residue on his hands, this diagram shows that for each correct positive result, there would also be two false positive results, where the test wrongly showed that the suspect had gunpowder residue on his hands. In a courtroom case, there could be advantages in treating this information step-by-step via several simpler diagrams—for instance, one diagram that deals only with true and false positives, and another that deals only with true and false negatives.

But either way, you can quickly look at this and see that the chances of this being a correct negative result are overwhelmingly high. The chart is practically all white. All other possibilities are mathematically trivial compared to the true negative.

Such a chart would be handy for showing the chances of a good outcome from undergoing a particular medical treatment, and how often you get a good outcome from *not* undergoing the treatment. If you saw this type of chart when the doctor gave you the good news, you probably would not fret about the slight chances on the fringes. The image would reinforce what the doctor was saying: that the chance of a bad outcome is statistically insignificant. You could also use such a chart to demonstrate the likelihood of winning the lottery or being struck by lightning.

We're not the first to try using a different visual representation to explain probabilities. As we've seen, researchers like Gigerenzer have been using visual representations for particular tasks. However, our way of working with information springs from a deeper underlying model. That extra depth gives a lot of extra power.

A key advantage of having an in-depth framework is that it provides systematic guidance on a whole batch of other ways of visually representing information. You have an entire toolbox, not just a single chisel or a single saw, and you have guidance on which tool to use for which purpose. This lets you display information in a way that's easier to understand and more likely to lead to the right conclusion. The cold case and serious crime police are keen on that, because it should reduce the time they waste by missing or misinterpreting key points. In a murder case, where the first forty-eight hours are crucial, that could make a big difference.

We're still in the early stages of this work, setting up the collaborations and the infrastructure. It's a time of changes for us, as new opportunities and new challenges open up before us.

What Happens Next

I have a fake crystal ball on my office desk. I bought it when I was doing my Voynich work, partly as a joking reference to the skrying crystal that Dee and Kelley used in their séances, and partly because I wondered whether it might give me some unexpected insight into how Dee and Kelley operated. (It didn't. . . .) If it did show the future, it would almost certainly show interesting times ahead for Verifier. Quite what form those interesting times will take, though, is harder to predict.

With the visualization tools we're building, there are a lot of possibilities. It may be that those tools will become part of the standard toolkit for working researchers and a standard part of how science is done. Perhaps physicians trying to calculate the statistical risks their patients face will routinely use one of those tools to visualize the risks and so avoid making disastrous medical decisions. Maybe researchers planning their next project will use one or more tools to map out systematically the possible questions to ask, so as to avoid missing some key point that's a blind spot for them.

The tools should also help researchers explain their work to the public. Increasingly, researchers are expected not only to make their data publicly available but also to explain their research to the public. They worry that their work will be misinterpreted because so many people—experts and lay public alike—have problems understanding scientific data. If they knew how to create compelling charts that exploited the mathematics of desire, they would be able to share their findings with confidence and know that their message was being received. If they knew how to spot logical fallacies, they might be better prepared to discuss their work with the media and on the Internet.

Researchers are already finding ways of using those tools that we hadn't envisaged. One colleague, for example, is a medical researcher. A key technique in medical research is the systematic review of published research findings to see just how well a given drug or treatment is working. Until now, those reviews have been limited to the published research in the reviewer's language, for obvious reasons. When our colleague realized that the Search Visualizer would let her identify relevant records in other languages, she immediately saw that this might be a game changer. She could now go beyond the literature in one language and start doing a comprehensive review of the literature from around the world. That significantly increases the chances of locating a key finding that might otherwise go unnoticed for years because of language barriers. Most researchers can describe how some key part of their discipline hit just that problem in the past, like Darwin being unaware of Mendel's discoveries about genetics and their immense implications for Darwin's work on evolution.

As more researchers start to appreciate these uses, the tools we're producing will spin off in directions that we can't predict. That's fine with us; the key thing is that the tools should be helping people. For our own core work on Verifier, the future is likely to be equally interesting and equally unpredictable.

There are a lot of significant problems still waiting to be tackled with the new, light version of Verifier. We'll probably need to develop new software for it, in addition to the visualization tools. Even with software support, the challenge won't be easy. We'll almost certainly make mistakes along the way, because people—including experts—do make mistakes. The key question is whether those mistakes are useful ones that help to identify dead ends and identify new approaches that might finally solve one of the big problems that affect people's lives. Making *that* type of mistake is something we could live with.

Acknowledgments

We'd like to thank everyone who helped bring this project to fruition. We would particularly like to thank our agent, the wonderful Yfat Reiss Gendell; also Cecilia Campbell-Westlind, Erica Walker, and the team at Foundry Literary + Media. At Harper-One, we owe a debt to our editor Roger Freet and his team: assistant editor Babette Dunkelgrun, production editor Lisa Zuniga, designer Ralph Fowler, copy editor Elizabeth Berg, and cover designer Noon, San Francisco.

High on the list to thank are Gordon's academic colleagues from the world of research: Jo Hyde, Marian Petre, Rich SantaColoma, René Zandbergen, and others too numerous to mention—they know who they are. Thanks are also due to Search Visualizer development colleagues Gerry Brennan, Ed de Quincey, David Miller, Julian Madle, Clyde Musgrave, Michael North, and everyone else involved in that story. Also Ian, Tom, Ash, and Stephen for their cheerful and efficient technical support.

And last, but very far from least, our long-suffering families.

Regardless of the assistance we've received from these fine friends and colleagues, any errors are entirely our own. Please direct questions and comments to our website at www.hydeandrugg.com.

Notes

Throughout this book, we've had to simplify topics to highlight the key points and keep the book to a reasonable length. The notes are intended as a starting point for more in-depth reading, if something we've written catches your interest. We hope you'll find them useful.

Chapter 1

De Groot, A. D. *Thought and Choice in Chess.* Mouton: The Hague, 1965. Groot's original book on chess experts.

Ericsson, K. A., N. Charness, P. J. Feltovich, and R. R. Hoffman, eds. *The Cambridge Handbook of Expertise and Expert Performance.* Cambridge: Cambridge University Press, 2006. A good place to start on expertise.

Perrow, C. *Normal Accidents: Living with High Risk Technologies.* Princeton, NJ: Princeton University Press, 1999. People tend to think of accidents as aberrations, but the full story is more complex. This is a good introduction to how accidents can be almost inevitable by-products of systems.

Reason, James. *Human Error.* Cambridge: Cambridge University Press, 1990. A good introduction to regularities in human error, including frequency capture errors.

Rugg, Gordon, and Marian Petre. *A Gentle Guide to Research Methods.* Maidenhead: Open University Press/McGraw-Hill, 2006. This contains chapters about the main elicitation techniques, including card sorts and laddering, with worked examples and handy hints. It also has a section on knowledge representation.

If you're interested in expert systems, any good textbook on artificial intelligence will include an overview of the topic. This is the best place to start for most readers. Books specifically about expert systems tend to be written for specialists and are highly technical.

Chapter 2

Baddeley, Alan. *Human Memory: Theory and Practice*. Hove: Lawrence Erlbaum, 1990. A good introduction to the topic of human memory.

Kahneman, D. P. Slovic, and A. Tversky. *Judgment Under Uncertainty: Heuristics and Biases*. New York: Cambridge University Press, 1982. The classic early text for judgment and decision making.

Loftus, E. F., and J. C. Palmer. "Reconstruction of Automobile Destruction: An Example of the Interaction Between Language and Memory." *Journal of Verbal Learning and Verbal Behaviour* 13 (1974): 585–89. The classic paper on memory distortion.

Maiden, N. A. M., and Gordon Rugg. "ACRE: A Framework for Acquisition of Requirements." *Software Engineering Journal* 11, no. 3 (1996): 183–92. The ACRE paper is available online in various places. The core concepts are summarized in Rugg and Petre, *A Gentle Guide to Research Methods*.

Miller, G. A. "The Magical Number Seven, Plus or Minus Two: Some Limits on Our Capacity for Processing Information." *Psychological Review* 63 (1956): 81–93. The seminal paper on short-term memory.

Sommerville, Ian, and Pete Sawyer. *Requirements Engineering: A Good Practice Guide*. Hoboken, NJ: Wiley, 1997. If you're interested in requirements engineering, this is a good place to start.

Chapter 3

Gigerenzer, Gerd. *Gut Feelings: Short Cuts to Better Decision Making*. Harmondsworth: Penguin, 2008. A very readable introduction to Gigerenzer's work.

Tufte, E. R. *The Visual Display of Quantitative Information*. 2nd ed. Cheshire, CT: Graphics Press, 2001. Tufte's work, particularly this one, provides a classic overview of diagrams and visual representation.

Wright, G., and Peter Ayton, eds. *Subjective Probability*. Chichester: Wiley, 1994. A thorough introduction to the work of Peter Ayton, Gerd Gigerenzer, and others who follow the same approach.

Chapter 4

Axelrod, R. M. *Structure of Decision: The Cognitive Maps of Political Elites*. Princeton, NJ: Princeton University Press, 1976. The source of the cognitive causal maps approach described in this chapter.

Bem, Sandra L. "The Measurement of Psychological Androgyny." *Journal of Consulting and Clinical Psychology* 42 (1974): 155–62. The description of androgyny theory in this chapter is based largely on Sandra Bem's work.

Euler, L. "Solutio problematis ad geometriam situs pertinentis." *Commentarii academiae scientiarum petropolitanae* 8 (1741): 128–40, http://math.dartmouth .edu/~euler/docs/originals/E053.pdf. Euler's original paper on graph theory (in Latin) is available online in facsimile form.

Gerrard, S., and J. Dickinson. "Women's Working Wardrobes: A Study Using Card Sorts." *Expert Systems* 22, no. 3 (2005): 108–14.

Kuhn, T. S. *The Structure of Scientific Revolutions.* 3rd ed. Chicago: University of Chicago Press, 1996. Kuhn's book on paradigms.

Likert, R. (1932). "A Technique for the Measurement of Attitudes." *Archives of Psychology* 140 (1932): 1–55. The original paper on Likert scales.

Ore, O. *Graphs and Their Uses.* Washington, DC: Mathematical Association of America, 1996. A good modern introduction to graph theory.

Rugg, Gordon. "Quantifying Technological Innovation." *Palaeoanthropology* 2011 (2011): 155–66. My paper on measuring complexity.

For visualizing online search, the Search Visualizer website is http://www .searchvisualizer.com. There's also a blog containing articles about ways of using the Search Visualizer at searchvisualizer.wordpress.com.

I should probably note that if you look into the scientific literature about stone age tools, you'll see several spellings for the same thing: *handaxe, handax, hand axe,* and *hand ax.* As if that wasn't confusing enough, modern dictionaries say that those same spellings can be used to refer to two different things. One is also known as a "hatchet": a small ax, with a handle and a head, designed to be used with one hand. The other is a stone age tool, oval in shape, designed to be used without a handle, like the one shown in Figure 6. Throughout this book, we're only talking about that stone age tool.

Chapter 5

McManus, I. C. "Scrotal Asymmetry in Man and in Ancient Sculpture." *Nature* 259 (1976): 426, doi:10.1038/259426b0.

Preece, J., Y. Rogers, and H. Sharpe. *Interaction Design: Beyond Human-Computer Interaction.* Chichester: Wiley, 2002. A good introduction to software interface design.

Rugg, G., and M. Mullane. "Inferring Handedness from Lithic Evidence," *Laterality* 6, no. 3 (2001): 247–60.

Vickery, B. C. *Faceted Classification: A Guide to the Construction and Use of Special Schemes*. London: Aslib, 1960. The classic introduction to facet theory in librarianship.

A note for military readers: We've simplified the story of Cannae down to its key points for clarity and brevity. For example, we didn't mention that some Romans managed to escape Hannibal's trap before it completely closed; one of those survivors was a soldier called Publius Cornelius Scipio, who went on to defeat Hannibal at the battle of Zama, finally winning the war for Rome.

Chapter 6

D'Imperio, M. E. *The Voynich Manuscript: An Elegant Enigma*. Laguna Hills, CA: Aegean Park Press, 1978. One of the earliest books about the Voynich Manuscript, whose title inspired the title for my *Cryptologia* article.

Woolley, B. *The Queen's Conjurer: The Science and Magic of Dr. John Dee*. New York: Henry Holt, 2002. A good recent biography of Dee.

The Voynich Manuscript section of the Beinecke library website is at http://beinecke.library.yale.edu/digitallibrary/voynich.html.

Several websites are key resources for Voynich researchers:

René Zandbergen's website: http://www.voynich.nu/

Jorge Stolfi's website: http://www.dcc.unicamp.br/~stolfi/voynich/

Philip Neal's website: http://www.voynichcentral.com/users/philipneal/

There's material about my own Voynich work at http://www.scm.keele.ac.uk/staff/g_rugg/voynich/moreIllustrations.html.

My Penitentia Manuscript can be found at www.scm.keele.ac.uk/research/knowledge_modelling/km/people/gordon_rugg/cryptography/penitentia/index.html. As far as I know, it's the longest uncracked ciphertext on the Internet. My Ricardus Manuscript (another long, uncracked ciphertext) can be found at www.scm.keele.ac.uk/research/knowledge_modelling/km/people/gordon_rugg/cryptography/ricardus_manuscript.html.

Chapter 7

D'Agnese, Joseph. "Scientific Method Man: Gordon Rugg Cracked the 400-Year-Old Mystery of the Voynich Manuscript." *Wired* 12, no. 9 (September 2004), www.wired.com/wired/archive/12.09/rugg.html.

Rugg, Gordon. "An Elegant Hoax? A Possible Solution to the Voynich Manuscript." *Cryptologia* 28, no. 1 (January 2004): 31–46.

Rugg, Gordon. "The Mystery of the Voynich Manuscript." *Scientific American* (July 2004): 104–9, www.scientificamerican.com/article.cfm?id=the-mystery-of-the-voynic-2004-07.

Schinner, Andreas. "The Voynich Manuscript: Evidence of the Hoax Hypothesis." *Cryptologia* 31, no. 2 (2007): 95–107.

Chapter 8

Blakeslee, S., and V. S. Ramachandran. *Phantoms in the Brain*. London: Fourth Estate, 1999. A good introduction to Ramachandran's neurological work.

Borgia, G. "Experimental Blocking of UV Reflectance Does Not Influence Use of Off-Body Display Elements by Satin Bowerbirds." *Behaviour Ecology Advance Access* (2008), doi: 10.1093/beheco/arn010.

Darwin, Charles. *The Origin of Species*. 2nd ed. Oxford: Oxford University Press, 1996 [1860]. Darwin's work is still highly readable. The second edition is generally viewed as better produced than the first edition.

Elman, J., E. Bates, M. H. Johnson, A. Karmiloff-Smith, D. Parisi, and K. Plunkett. *Rethinking Innateness*. Cambridge, MA: MIT Press, 1996. This interesting source on robots, instinct, and innateness reexamines concepts like *innate* and *instinctive* in terms of the insights offered by approaches such as artificial neural networks.

Hirstein, W., ed. *Confabulation: Views from Neuroscience, Psychiatry, Psychology and Philosophy*. Oxford: Oxford University Press, 2009. A comprehensive overview on confabulation.

Madden, J. R., and A. Balmford. "Spotted Bowerbirds *Chlamydera Maculata* Do Not Prefer Rare or Costly Bower Decorations." *Behavioral Ecology and Sociobiology* 55 (2004): 589–95.

Maynard Smith, J. *Evolution and the Theory of Games*. Cambridge: Cambridge University Press, 1982. On Johnson-Laird's game theory.

Ramachandran, V. S., and W. Hirstein. "The Science of Art: A Neurological Theory of Aesthetic Experience." *Journal of Consciousness Studies* 6 (1999): 15–51.

Rodriguez, I., A. Gumbert, N. H. de Ibarra, J. Kunze, and M. Giurfa. "Symmetry Is in the Eye of the 'Beeholder': Innate Preference for Bilateral Symmetry in Flower-Naïve Bumblebees." *Naturwissenschaften* 91 (2004): 374–77. On insects' preference for symmetry.

Winkielman, P., J. Halberstadt, T. Fazendeiro, and S. Catty. "Prototypes Are Attractive Because They Are Easy on the Mind." *Psychological Science* 17, no. 9 (2006): 799–806. A good example of the numerous papers about prototypical images and attractiveness.

Chapter 9

Asperger, Hans. "Autistic Psychopathy in Childhood," in *Autism and Asperger Syndrome,* translated and edited by U. Frith. Cambridge: Cambridge University Press, 1991.

Gerrard, Sue, and Gordon Rugg. "Sensory Impairments and Autism: A Re-Examination of Causal Modelling." *Journal of Autism and Developmental Disorders* 39, no. 10 (2009): 1449–63.

Kanner, L. "Autistic Disturbances of Affective Contact." *Nervous Child* 2 (1943): 217–50. Kanner's original paper is much mentioned but probably not often read.

Ramsay, A. M. *Myalgic Encephalomyelitis and Post Viral Fatigue States: The Saga of Royal Free Disease.* 2nd ed. Edited by Marion Jowett. New York: Gower Medical Publishing, 2005. An interesting book on early research into ME/CFS. Ramsay views ME solidly within the virological tradition, as most likely to be a postviral condition. It was printed by the ME Association and can be obtained from them.

Snowling, M. J., and C. Hulme, eds. *The Science of Reading.* Oxford: Blackwell Publishing, 2005. A comprehensive overview of psychological research into reading. It's not a light read, but it's thorough and authoritative.

Williams, Donna. *Nobody Nowhere: The Extraordinary Autobiography of an Autistic.* New York: Times Books, 1992. An insider's perspective into autism, with some fascinating insights.

Williams, Donna. *Somebody Somewhere: Breaking Free from the World of Autism.* New York: Three Rivers Press, 1994.

Chapter 10

For a good overview of the use of old vellum in antiquity, Rich SantaColoma's website is a comprehensive resource: http://proto57.wordpress.com/.

Index

Page numbers in *italics* refer to illustrations.

ACRE framework, 38, 40, 49–51, 19–21, 147
aesthetics, 179–207; consumer research and, 201; entertainment and, 201–2; gender politics and, 200–201; mate selection, 186–88, 195–96; mathematics of desire, 179–207; novelty and, 198–200; sales and, 203–4; speculations and, 200–207; superstimuli and, 189–96; symmetry and, 183–86, 191–92; uncanny valley, 192–94, 206; underlying principles of, 179–207; warfare and, 204–6
affordances, 39–40
Africa, 109, 137
AIDS, 241
aircraft, 31, 42–43, 46, 47, 183, 268; design, 184–86, 192
air traffic, 82
alarm bells, mental, 165–66
alchemy, 126, 127, 128, 146, 156
Alexander the Great, 204
algorithms, 88–94, 166; genetic, 185
Alien (film), 185–86
alphabets, 133; Voynichese, 133, *134*, 148
already-solved issue, 260
Alzheimer's, 103, 123–25, 243
Amazon, 200
American Psychiatric Association, 224
androgyny, 76, 271–72
angels, 141–46, 255
animals, 184, 187, 189, 194–95, 239

annihilation, battle of, 115
anthropology, 80–87, 192
antler, 84, *84*, 85
aorta, 239
Apple, 79
Arabic, 137
archaeology, 23, 83–87, 101–2, 147
architecture, 191–92
arcs, 82–87
Arctic, 33
Aristotle, 117
arms races, 188–94
arrow, paradox of, 52–53
art, 96–98, 182, 183, 185, 189–90, 199, 201, 202
artificial intelligence (AI), 20, 24, 182, 185, 196–98
artificial neural networks (ANNs), 197
Asperger, Hans, 210–12, 236
Asperger's syndrome, 210–11
astrology, 6, 127
astronomy, 54
asymmetry, 184, 185, 186, 192, 257; warfare, 205–6
Atlantis, 125
ATM, 69–71
attractiveness, 74, 110; gender politics and, 200–201; mate selection, 186–88, 195–96; perceptions of, 74–78; superstimuli and, 189–96; symmetry and, 183–86, 191–92, 195; website design and, 74–78

auditory problems, 220–229
Australia, 194
Austria, 176, 210
autism, 167, 207, 209–42, 243–44;
    definitions, 210–12, 217–22, 225–26;
    diagnosis, 210, 213–14, 217–22,
    223–25; epidemic, 259; findings
    and publication, 232–38; groups
    versus individuals, 217–22; medi-
    cal models, 212–14; spectrum, 212,
    213; syndrome, 212, 213, 225; triad
    of impairments, 212, 213; types
    of behavior impaired in, 214, 216,
    217–22, 225–27, 236; Verifier analysis
    of, 209–42, 257–67
axes, complexity of, 79–87
Ayton, Peter, 59–64

bacteria, 7–8, 222, 238–39
Baddeley, Alan, 41
balance, 184–85
Balmford, Andrew, 196
bank tellers, 56, 57, 62, 63
Bantu, 137
basketball, 67
Basque, 137–38
Belgium, 205
benign myalgic encephalomyelitis
    (ME), 230
Berlin Wall, fall of, 174
Bettelheim, Bruno, *The Empty Fortress:
    Infantile Autism and the Birth of Self*,
    211–12
biases, 58, 118, 206, 235
Bible, 142
Bigfoot, 125
big picture, 264–65
binary reasoning, 73
binomial distribution, 163
Blandford, Ann, 118
Bleuler, Paul Eugen, 211
blogs, 170, 245
blood transfusions, 66
blood vessels, 239–40
body language, 49
boredom, 198
Borgia, Gerald, 196
bowerbirds, 194–98
bradycardia, 92

brain, 4, 26, 27, 37, 40, 41, 45, 59,
    179–80, 184, 190, 197, 198, 202, 206,
    240; dyslexia and, 227–30
bridges, 81–82, 96
Bronze Age technology, 83–87
Brumbaugh, Robert S., 167
bulk carrier ships, 32–35
Burton, Mike, 21

calligraphy, 112, 149, 156, 160–61,
    246–48
cancer, 63, 103
Cannae, Battle of, 113–15, 204–6, 235
carbon 14 dating, 244, 274; on Voynich
    Manuscript, 244–57
Cardan grille, 150–52, 164, 250–51, 255
Cardano, Gerolamo, 150, 164, 253, 255
card sorts, 22–23, 71–74, 203, 242; fash-
    ion and, 71–74
cars, 44–45, 168–69, 188, 192, 199;
    hybrid, 193
cartography, 163
cartouche, 254
categorization, 22–23, 51–52, 72–74,
    236–42, 261; medicine, 238–42
cause-and-effect theories, 261–64
cavitation bubbles, 174
Celtic Gaul, 113, 233
Centers for Disease Control and Pre-
    vention, 231
Chalmers, John, 240–42
Champollion, Jean-François, 254
chance, 67
chemistry, 216, 242
chess, 14–15, 16, 181, 190
Chittenden Manuscript, 249
chronic fatigue syndrome, 227, 230–32;
    Verifier analysis of, 230–32, 267
City University London, 29
Civil War, 115
Clinton, Bill, 44
Clinton, Sir Henry, 151
closure, 206–7
codes, 127–28, 130; breaking, 2, 3,
    130–35, 140, 157, 163, 168, 257;
    D'Agapeyeff Cipher, 130–32, 163–64;
    Internet, 129; Penitentia Manuscript,
    132, *132*, 133; Voynich Manuscript
    and, 127–29, 133–35, 140, 151–52,

166–68, 170, 171, 172, 175, 178, 252–54, 256, 257

Cold War, 173–74

Coleman, Mary, 222

color, 99, 182; coding, 92

community, 170–71, 237

compiled skills, 45–46

complexity, 79–87, 170, 199; of technology, 79–87

computers, 4, 16, 25–27, 190, 267; tablet, 193; visualizing online searches, 88–94, *90–93;* website design, 74–78. *See also* Internet; software

computer science, 196–98

concepts, 264

cone of percussion, 104–5, 106, 108, 109, 110

confabulation, 190–91

confirmation bias, 235

conjunction fallacy, 55–58, 62, 63

consumer, 71, 199–200, 203; novelty and, 198–200; research, 201; sales, 203–4; website design and, 74–78

consumer electronics, 193

container shipping, 32–35

cosmetics, 189

Craig, Daniel, 187

cross-disciplinary research, 233–35

cryptography, 128–35, 140–41, 146, 149, 163, 166–68, 253–55

Cryptologia, 167–68, 175

Currier, Captain Prescott, 251

Currier Y, 251–52, *252,* 253

D'Agapeyeff, Alexander, 131, 163–64

D'Agapeyeff Cipher, 130–32, 163–64

Darius, emperor, 204

Darwin, Charles, 187, 279

dead end, 261

decision making, 58–59, 67

Decision Support Systems (DSSs), 270

Dee, John, 141–46, 149, 156–57, 168, 255, 256, 278

deep structure, 119–20

de Groot, Adriaan, 14

design, 74, 181; Golden Ratio, 180–83; nonverbal signage, 96–99; novelty and, 198–200; symmetry and, 183–86, 191–92; website, 74–78

desire, 179–207; consumer research and, 201; entertainment and, 201–2; gender politics and, 200–201; mate selection, 186–88, 195–96; mathematics of, 179–207, 279; novelty and, 198–200; sales and, 203–4; speculations and, 200–207; superstimuli and, 189–96; warfare and, 204–6

Diagnostic and Statistical Manual of Mental Disorders (DSM), 210, 212, 224, 225, 228

diagrams, 68, 69–99, 242, 261, 264, 274–77; complexity and, 79–87; graph theory, 81–87; handedness, 110–11, *111,* 112; scales, 75–78; signage, 94–99; Venn, 57, *57;* visualizing online searches, 88–94, *90–93*

dialects, 139, 162

dice, 149, 150

discouragement, 77–78

division, 60–61

DNA, 239

domain experts, 19

dopamine, 262

double envelopment, 115

Down syndrome, 240

driving, 44–45, 168–69

Duffee, Boyd, 265

dyslexia, 167, 227–30, 237; Verifier analysis of, 227–30, 237, 267

early humans, 83–87; handedness, 102, 105–10; technology, 83–87

economics, 71, 195–96, 199, 200, 201

education, 5

Egyptian hieroglyphs, 135, 254

Einstein, Albert, 54

elicitation, 5, 50, 120, 147

Elizabeth I, Queen of England, 3, 6, 141, 255

encouragement, 77–78

England, 18, 51, 73, 86, 146, 151, 184, 230

English, 136, 137, 138, 140, 154, 155, 159

Enochian language, 139–40, 145, 171, 255–56

entertainment, 201–2, 206; industry, 198, 199, 200; preferences, 201–2

episodic memory, 40–41
erasures, 177–78, 244
errors, 8–9, 12, 17, 27, 30–36, 47, 99,
  103, 118, 280; ambush in your mind,
  101–21; desire and, 179–207; erasures,
  177–78, 244; expert, 12–14, 30–36,
  47, 49–68, 101–21, 243; frequency
  capture, 31, 46; inbuilt human preferences, 179–207; logical fallacies and,
  51–64; of omission, 66; probability
  and, 59–68; Verifier Method for, *see*
  Verifier Method; Voynich Manuscript
  and, 177–78
Estonia, 137
Euler, Leonhard, 81–82, 96
evidence, 273–78; forensic, 273–78; key
  evidence chains, 261–64
evolution, 187, 279
evolutionary ecology, 117, 119, 187, 188,
  194, 201
experts, 4, 5, 11–27, 118, 123, 191, 243;
  card sorts, 22–23, 71–74; domain,
  19; errors, 12–14, 30–36, 47, 49–68,
  101–21, 243; forensic, 273–78; gaps,
  261, 265; imperfect, 49–68; interviewing, 18–21; knowledge, 11–27, 29–47,
  49–68; laddering, 23, 214–16; mind, 5,
  11–27, 29–47; probability and, 59–68;
  problems that science tackles and
  Verifier Method, 257–66; think-aloud
  technique, 21–22; types of, 54–55;
  Voynich Manuscript and, 168–78
explicit knowledge, 40–47
eye contact, and autism, 220, 221

fabricatory depth, 83
facial proportions, 183–84
fallacies, logical, 51–64
false positives and negatives, 275–76,
  276, 277
fashion, 71–72, 73, 188, 189, 192
femininity, 76–77, 187, 271–72
fencing, 13
fetishes, 194
Finland, 137
Finno-Ugric, 137
flashbulb memory, 40–41
flint, 83, *84*, 85, 103, 104, 105, 106,
  108, 110

Flournoy, Théodore, 139
folklore, 192–93
forensics, 273–78
France, 95, 138, 205
French, 140
frequencies, 63–64, 163, 274
frequency capture errors, 31, 46
future, 42
future requirements, 39–40

game theory, 117–20, 186–88, 264
gaps, expertise, 261, 265
gastritis, 8
Gaugamela, Battle of, 204–5
gender, 71–74, 76, 271–72; androgyny
  theory, 76, 271–72; fashion and,
  71–73; mate selection, 186–88,
  195–96; perceptions of attractiveness,
  76–77; politics, 200–201
genetic algorithms, 185
genetics, 65, 188, 236, 239, 240, 279
Gentle, Mary, *Rats and Gargoyles*, 181
geology, 20–21, 23, 24, 44, 67, 216
German, 136, 137
Germany, 59–61, 92, 205, 233
germ theory, 238
Gerrard, Sue, 71–74, 78, 178, 209–10,
  217, 219–20, 225, 240, 265, 266
Giger, H. R., 185–86
Gigerenzer, Gerd, 59–64, 71, 275
GIGO, 117–20
Gillberg, Christopher, 222
Glass, Philip, 203–4
global economy, 32, 80
Go, 16
Golden Ratio, 180–83
Google, 74–75, 79, 199
Gould, Judith, 212
graph theory, 81–87
Great Exhibition (1851), 11
Greece, ancient, 52–53, 180–81
Greek, 136
Greenpeace, 56–58, 63
Grice, Paul, 43
grilles, 154; Cardan, 150–52, 164,
  250–51, 255; Voynichese, 154–67,
  250–55
gut feel, 37
Gut Feel, 64

hammerstone, 83, 84, *84*, 85, 103, 104, 105
hand ax, 83, 84, *84*, 85, 86–87, 101–2
handedness, human, 102, 105–113, 116, 118, 273; diagrams, 110–11, *111*, 112; early, 102, 105–10
handwriting, and carbon dating, 248–50
Hannibal, 114, 115, 175, 204, 233
hard-wired preferences, 196–98, 200
heart, 239
Hebrew, 137
heightism, 201
Helicobacter pylori, 7, 8
herbalism, 126, 127
heuristics and biases, 58, 62, 64, 107, 195
hieroglyphs, 135, 254
Hirstein, William, 189–90
hoaxing, 265; Voynich Manuscript and, 126, 129, 130, 138–52, 154, 156–57, 166–67, 168, 172, 175–78, 179, 207, 244–45, 250–57, 265–67
Holland, Nikki, 199
honest dice, 149
horror movies, 192
hot hands, 67
human error. *See* errors
humanist hand, 133, 248
Hungary, 137
hybrids, 193
Hyde, Jo, 118–21, 123, 171, 257, 258, 268

implicit learning, 46
inbuilt human preferences, 179–207
independence, 64–65
Industrial Revolution, 11
innovation, and complexity, 79–87
insects, 184; locomotion, 197–98
instincts, 196–98
interdisciplinary research, 233–35
International Classification of Diseases (ICD), 210, 212
international trade, 128–29
Internet, 2, 52, 81, 82, 279; codes, 129; traffic, 82, 83; website design, 74–78
interviews, 18–21, 39, 72; structured versus unstructured, 19–20
inverse frequency weighting, 198–99
IQ, 218

Italy, 1, 11, 52, 113–15, 126, 128, 255

Japan, 2, 81, 183; robotics, 192
Johns Hopkins University Hospital, 210
Johnson-Laird, Philip, 117, 119
journals, academic, 167–68, 236–37, 244, 258, 265, 279
judgment and decision making, 58–59, 67

Kahneman, Daniel, 17, 56–58
Kanner, Leo, 210–12, 219, 221, 225, 235, 236
Keele University, 36, 38, 147
Kelley, Edward, 139, 141–47, 149, 156–57, 168, 244, 245, 250, 255–56, 278
key themes and evidence chains, 261–64
Kinsey, Alfred, 200
knowledge, 5–6, 36; ambush in your mind, 101–21; categories of, 36–47; cycle, *5;* desire and, 179–207; diagrams and, 68, 69–99; expert, 11–27, 29–47, 49–68; explicit, 40–47; future requirements, 39–40; hierarchies, 23; imperfect expert and, 49–68; inbuilt human preferences, 179–207; nonverbal, 16, 68, 69–99; semitacit, 41–44; tacit, 44–47; taken-for-granted, 43–44; Verifier Method and, *see* Verifier Method; within, 29–47
knowledge acquisition bottleneck, 18

laddering, 23, 214–16, 235
language, 68, 92–94, 109, 134–35; autism and, 211, 218, 220; families, 136, 137; hoaxes, 139–41; short words, 136–37; unidentified, 127, 129–30, 138–41, 168, 170, 175, 252, 254, 257; Voynich Manuscript, 129–30, 133–52, 154–68, 175, 248–57
Laterality, 108
Latin, 136, 137, 149
left-handedness, 102, 105–13, 273
Likert, Rensis, 75
Likert Scales, 75–76
Linda problem, 55–58, 62, 63, 64, 71
literature review, 261
loaded dice, 149

Loftus, Elizabeth, 40–41
logic, 13, 30, 51–55, 117–20, 279;
    formal, 117–20; logical fallacies and
    human error, 51–64; types of, 120
"logic fail" example, 117
logistics, 113–15
London, 18, 51, 151, 230, 256
long-term memory, 42, 190
Lorenz, Konrad, 189
Lucas, George, 13

mad cow disease, 241
Madden, Joah, 196
Maiden, Neil, 31, 36, 49–51
malaria, 167
manufacturing, 112, 260
maps, 163, 264
marketing, 112, 182
marriage, 72
Marshall, Barry, 7–8
Martians, 139–40, 171
Martine, Giselle, 203
masculinity, 76–77, 187, 271–72
mate selection, 186–88, 195–96
mathematics, 3, 52, 53, 60–61, 62, 117,
    178; of desire, 179–207, 279; game
    theory, 117–20, 186–88; Golden
    Ratio, 180–83; graph theory, 81–87;
    set theory, 52
measurement theory, 76, 235
media, 3, 168, 193, 222, 223, 245, 279
medicine, 7–8, 17–18, 63, 64, 66, 112,
    123–25, 191, 193, 200, 274, 278, 279;
    categorization and, 238–42; "paper
    studies," 66; software-supported,
    17–18; statistics and, 63, 64, 65, 66;
    Verifier Method and, 209–42, 257–66
melanopsin, 262
melatonin, 262
memory, 15, 36–38, 59, 97; episodic,
    40–41; explicit knowledge and, 40–47;
    flashbulb, 40–41; long-term, 42, 190;
    semantic, 40; short-term, 42–43,
    60–61, 190, 202; types of, 37–47
Mendel, Gregor, 279
Mendeleev, Dmitri, 242
mental disorders, 210; autism and Veri-
    fier Method, 209–42; classification of,
    225–27

Middle Ages, 52, 126, 127, 142, 181
Middlesex University, 117
military, 12–13, 14, 33, 113–15, 173–74,
    204–6
mini-roundabouts, 95–96
mistakes. See errors
Model 202 Boomerang, 185
Mondrian, Piet, 182
Mori, Masahiro, 192
motion, 54
Mozart, Wolfgang Amadeus, 15
Mullane, Maureen, 108
music, 15, 116, 180, 199, 201, 202, 206;
    sales, 203–4
MYCIN, 17–18

Napoleon Bonaparte, 115
NASA, 240, 242
NATO, 173, 174
Nazism, 2
Neal, Philip, 151, 171
Neanderthals, 109
negatives, false, 275–76, 276, 277
Neisser, Ulric, 40
Neolithic Age, 87
Netflix, 200
neurophysiology, 38, 45, 67, 216,
    233–34, 264
Newbold, William R., 167
Newton, Isaac, 54
9/11, 40
nodes, 82–87
nonverbal knowledge, 16, 68, 69–99;
    diagrams, 68, 69–99; signage design,
    96–99
North Africa, 113
Nottingham, 18, 20, 21, 24, 29, 44, 72
novelty, 198–200, 201, 205

oak gall ink, 247–48
observation, 35–36
online searches, visualizing, 88–94,
    90–93
Open University, 73
Oswalt, Wendell, 80

paired scales, 76–78
paradox, 52–53
parallel processing, 25–26, 89–94, 270

paraphilia, 194
parchment, 245–47
Paris, 95
Parkinson's disease, 262, 263
Parthenon, 180–81
pattern matching, 24–27, 44, 71, 86, 87, 88–94, 99, 169, 202, 259–61, 266–67, 270, 274; diagrams, 69–99
pattern-welded swords, 147
Pattinson, Robert, 187
peacocks, 187–88
Pearl Harbor, Japanese attack on, 40–41
Penitentia Manuscript, 132, *132*, 133
periodic table, 242
Persia, 204
pharmaceutical companies, 124, 258
Philidor, François-André Danican, 14
physics, 54, 175
plants, 126, 127, 130–31, *131*, 132, 133, 135, 250, *250*, 253
Plato, 125, 184
playing card stencil, 251, 253
poker, 186
politics, 52, 188, 202–3, 222, 223; gender, 200–201; preferences, 202–3
Portland Vase, 11–12
positives, false, 275–76, *276*, 277
postviral fatigue syndrome (PVFS), 230
predators, 187, 188
predictions, 60–64, 199
preferences, inbuilt, 179–207
prefixes, 149, 154–67; Voynichese, 154–67, 253, 256
Price, Linda, 73
prions, 241–42
probability, 59–68, 273–78
problem-solving, 4–10
proper nouns, 254
proportions, 179, 180–83; Golden Ratio, 180–83; of human face, 183–84
prototypicality, 199, 268–69
psychology, 20, 36, 38, 67
pub darts, 112

questionnaires, 39, 72

Ramachandran, V. S., 189–90; *Phantoms of the Mind*, 224
randomness, 148–50, 164

ratios, 178, 180–83; Golden Ratio, 180–83
reading disorders, 229
ready-made solution, 260
real estate, 200, 203
realism, 12–14
Reason, Jim, 31
reasoning, 30–31; binary, 73; faulty, 30–36, 51–55; logical fallacies and human error, 51–64
Red Army, 233
refrigerator mothers, 211–12
Reichgelt, Han, 21
reification, 225, 229
religion, 188
representation, 5, 261, 264, 271, 273–78. *See also* diagrams
requirements, future, 39–40
requirements engineering, 29–30, 38, 43
rhetoric, 52
Richard III, King of England, 184
right-handedness, 102, 105–13
roads, 82; signage, 95–99; traffic, 82
robotics, 192, 193, 196–98
Rome, 126; ancient, 11, 52, 113–15, 116, 204–6, 235
roots, 149, 154–67; Voynichese, 154–67, 253, 256
Rosetta stone, 254
Rudolph II, Holy Roman emperor, 145, 146, 157, 255, 256, 257
rules of thumb, 58–59, 61, 107
Rutan, Burt, 184–85

sabers, 13
Sacks, Oliver, 224
sales, 203–4
SantaColoma, Rich, 246, 249, 256
Santorini, 125
Saul, Barnabas, 144, 145
scales, 75–78, 110; Likert, 75–76; paired, 76–78
Schinner, Andreas, 175, 244
schizophrenia, 211
Schlieffen Plan, 205
Schwarzkopf, Norman, 115
sculpture, 96–98
searches, online, 88–94
Search Visualizer, 7, 268–73, 279

Segways, 193
semantic memory, 40
semitacit knowledge, 41–44
Semitic language, 137
sequential processing, 26, 202
serotonin, 262
set theory, 52
sex, 71, 72, 271; gender politics and, 200–201; mate selection, 186–88, 195–96; preferences, 200; uncanny valley and, 193–94
Shadbolt, Nigel, 21
Shakespeare, William, 6
shell shock, 224, 232
short-term memory, 42–43, 60–61, 190, 202
signage, 94–99; nonverbal design, 96–99
skills, 36–38, 190; compiled, 45–46; types of, 38–47
skryers, 143–46
Slovic, Paul, 17, 56–58
Smith, Helene, 139–40
soccer, 59–64, 222–23, 259
sociology, 192, 200, 223
soft hammer, 84, *84*, 85
software, 4, 5, 7, 17–19, 25–26, 30, 34, 35, 42, 54, 182–83, 185, 202; bugs, 240; online search, 88–94; Search Visualizer, 7, 268–73, 279; short-term memory and, 42, 43; Verifier and, 268–73, 279–80; waterfall model, 18, 39
Somme, 115
Soviet Union, 1, 173–75, 233
Spain, 113, 138
speaking in tongues, 171
spearheads, 80–81
sports, 45, 59–64, 67, 97, 201, 202, 206, 222–23
sports psychology, 67
Stanford University, 17
Star Wars (film), 13
statins, 66
statistics, 62–68, 139, 269, 278
Stolfi, Jorge, 148–50, 167, 170, 171
stomach ulcers, 7–8
Stone, Chris, 168
Stone Age Institute, Bloomington, Indiana, 106

stone tools, 83–84, *84*, 85–87, 101–2, 103–4, *104*, 105–10
stress, 7
striking platform, 104–5
Strong, Leonell C., 167
Sudden Infant Death Syndrome (SIDS), 62, 65
suffixes, 149, 154–67; Voynichese, 154–67, 253, 256
Sulzer, Andreas, 176
supernatural, 143–46
superstimuli, 189–96
survival, 195
Switzerland, 81
syllables, 133, 138; Voynichese, 138, 148–52, 154–67, 253, 256
symmetry, 183–86, 189, 191–92, 195, 199, 205; warfare, 205–6
Systematic Literature Review, 261
Szymanski, Zoe, 75–78, 110

tables and grilles, 154–55, *155*, 156–65, *164–65*, 166–67, 250–55
tacit knowledge, 44–47
tactics, 113–15
taken-for-granted knowledge, 43–44
technology, 2, 8, 12, 29–30, 79, 172, 173–75, 182–83, 269; Bronze Age, 83–87; complexity of, 79–87
techno-unit, 80–81
television, 201–2
testing, 5
themes, key, 261–64
think-aloud technique, 21–22
this and that, calculation of, 55–58, 62, 63, 64, 71
Tinbergen, Niko, 189
"tip of the tongue" effect, 264
Torah, 177
torpedoes, 33, 173–75
Toth, Nick, 106, 108, 110
tourism, 193
Turkey, 59–60
Turkish, 137
Tversky, Amos, 17, 56–58

ulcers, 7–8
uncanny valley, 192–94, 206
University of Arizona, 244, 245, 249

University of California-Los Angeles, 80
Uomini, Natalie, 108–9

vellum, and carbon 14 dating, 244–57
Venn diagram, 57, *57*
Verifier Method, 3–10, 30, 51, 68, 110,
116–21, 123–25, 146, 152, 167, 168–
75, 234, 241, 244, 257–66, 279–80;
autism and, 209–42, 257–67; birth of,
116–21; chronic fatigue syndrome
and, 227–30, 267; combinations used
for Voynich Manuscript, 168–75; dyslexia and, 227–30, 237, 267; medicine
and, 209–42, 257–66; problems that
science tackles and, 257–66; software
and, 268–73, 279–80; Voynich case
studies, 152, 153–78, 244–57. *See also
specific tools*
Vienna, 210
Vietnam War, 115
viruses, 222, 238, 239, 241–42
vision, and autism, 220–21
visualization, 6, 7, 270, 273–78
Voynich, Wilfrid, 1, 125–28, 257
Voynich Manuscript, 1–3, 27, 103,
125–41, 207, 278; alphabet, 133,
134, 148; Cardan grille and, 150–52,
164, 250–51, 255; carbon 14 dating
on, 244–57; cipher text explanation,
127–29, 133–35, 140, 151–52, 166–
68, 170, 171, 172, 175, 178, 252–54,
256, 257; community, 170–71; dating
of, 176, 244–57; discovery of, 126;
erasures and corrections, 177–78,
244; experts and, 168–78; handwriting, 248–50; hoax hypothesis, 126,
129, 130, 138–52, 154, 156–57,
166–67, 168, 172, 175–78, 179, 207,
244–45, 250–57, 265–67; illustrations,
126, 127, 128, *131*, *134*, 156, *250;*
language, 129–30, 133–52, 154–68,
175, 248–57; plant pages, 126, 127,

130–31, *131*, 132, 133, *134*, 135, 250,
250, 253; publication, 167–68, 170,
171; randomness and, 148–50, 164;
"sunflower" page, 130–31, *131*, 132,
133, *134*, 250, *250;* syllables, 138,
148–52, 154–67, 253, 256; tables
and grilles, 154–55, *155*, 156–65,
164–65, 166–67, 250–55; text, 126,
127, 128–34, *134*, 135–52, 154–67;
unidentified language explanation,
127, 129–30, 138–41, 168, 170, 175,
252, 254, 257; Verifier combinations
used for, 168–75; Verifier studies, 152,
153–78, 244–57

walkways, 98
warfare, 12–13, 14, 33, 113–15, 128,
173–74, 204–6, 233; desire and,
204–6; trench, 224
Warren, Robin, 7–8
watches, 79
waterfall model, 18, 39
website design, 74–78
Wedgwood, Josiah, 11
Wedgwood pottery, 11
wild man, 192
Williams syndrome, 240
Wing, Lorna, 212
World Health Organization, 210
World War I, 115, 128, 205, 224
World War II, 2, 33, 233, 257
Wright, George, 63

X Files, The (TV show), 118

Yale University, 1; Beinecke Library,
125, 248

Zandbergen, René, 167, 176–77
Zeno, 52, 53
Zeno's paradox, 52–53, 75
zodiac, 127

SEARCH VISUALIZER

Now online:
searchvisualizer.com

◂ ◂ ◂ ▸ ▸ ▸

To learn more, visit our blog at:
searchvisualizer.wordpress.com

◂ ◂ ◂ ▸ ▸ ▸

Worth more than
a thousand words . . .